Thomas E. Little

Management Systems in Education

THE PROFESSIONAL EDUCATION SERIES

Walter K. Beggs, *Editor*
Dean Emeritus
Teachers College
University of Nebraska

Royce H. Knapp, *Research Editor*
Regents Professor of Education
Teachers College
University of Nebraska

Management Systems in Education

by

PAUL A. MONTELLO

Associate Professor of Education
and Urban Life
Georgia State University

and

CHARLES A. WIMBERLY

Assistant Professor of Industrial
Engineering Technology
Southern Technical Institute

PROFESSIONAL EDUCATORS PUBLICATIONS, INC.
LINCOLN, NEBRASKA

Library of Congress Catalog Card No.: 74-83592

ISBN   0-88224-081-1

Contents

Illustrations

CHAPTER 4

Tables

Preface

This book was developed to provide educators with a brief review of many concepts and practices that are currently being explored and utilized in educational management systems. It was the intent of the authors to provide an explanation of these concepts in sufficient detail to challenge the most thoughtful readers to pursue the subject but in a style appropriate for persons who are beginners in educational management. The book should prove helpful to teachers and administrators who have a need for developing a general understanding of management systems. Additionally, the book is a most appropriate text for school-board members and lay citizens who are interested in the systems approach to managing the educational enterprise.

Chapters 1 and 2 present a discussion of many basic concepts of systems theory. They also provide an explanation of selected techniques of systems analysis. The techniques included in Chapter 2 were selected because they show the greatest promise for use in education or are already in use by educators. Examples of applications to educational settings are provided in the discussion of each systems-analysis technique. Chapter 3 presents a discussion of several comprehensive management systems, and several techniques of systems analysis are illustrated. The application of systems theory to an actual educational program is provided in Chapter 4 so that the reader can gain an understanding of how systems techniques are applied to educational endeavors. In the final chapter, the authors outline some of the types of systems activities that can be expected to emerge in education. A glossary is provided as a ready reference for many of the concepts and terms that are frequently used in educational management systems. Finally, there is a list of suggested readings at the end of each chapter, as well as a very selective annotated bibliography as a guide for further study.

CHAPTER 1

Systems in Education

What is a system? Why are educators and administrators so interested in systems analysis, systems management, and computers?

In our modern technological society, over the last few decades, man has managed some rather remarkable achievements—such accomplishments as worldwide air travel, space travel, and a telephone system that allows instant communication throughout the country. In general, these achievements have resulted from the *systems approach*. Consider the telephone. How is it that when one dials the appropriate combination of numbers he can talk almost immediately with another party thousands of miles away? Think of it this way—each individual telephone is linked with every other telephone to create a *system*. A system is composed of a multiplicity of components, tied together for some common purpose.

Another system which can be readily understood is home heating. There are several components involved, including the furnace, warm air ducts, cold air ducts, a fan to move the warm air, and a thermostat for detecting temperature changes. When the thermostat detects the room air as being too cold, the furnace is automatically turned on, and the fan forces warmed air into the room until the thermostat detects that the room is at the proper temperature. At this point, a signal from the thermostat turns the furnace off. The components of this simple system work together to provide climatic control for the home, with each component doing its part.

Nature is a *supersystem*. All elements in nature are the components, and many of them are individual systems in their own right. Scientists think of the human organism as a system, composed of thousands of subsystems, such as the nervous system, the digestive system, and the muscular system. In fact, any animal or plant can be considered a system and, perhaps, a subsystem of a larger system.

For the purposes of this book, a *system* is defined as a multiplicity of parts, elements, or components, which interact with one another and work together for some common purpose. The parts, elements, and components of a system are often systems unto themselves and may, therefore, be called *subsystems.* A large system comprising several smaller systems and sub-systems is called a *supersystem.*

EDUCATIONAL SYSTEMS

David E. Barbee has developed the concept of systems in education well. He stated:

In order to develop more efficient and effective systems for meeting educational needs, more than a new teaching methodology is required. . . . The systems approach is not a sequential series of steps; it is a dynamic, interactive process. Initial objectives are modified as a result of later analysis; constraints may be modified as a result of seeing their impact on the cost of the system; the proposed solutions will be modified as a result of trade-off studies; the entire system may be redesigned as a result of the operational evaluation. Hence, at every step of the way, the results are analyzed to verify or modify earlier decisions.[1]

In this context, educational systems and systems technology take on a special meaning. "Technology, whether applied to education or sending men to the moon, is a process. It has no content; it is not the product that is derived by the process."[2]

J. Alan Thomas has stated that "educational systems are characterized by a sameness of organization and an apparent belief that there is a single 'best' method of combining resources for producing changes in the behavior of students."[3] Education is further characterized as a social institution functioning in a world which is in a constant state of change, and it therefore seems obvious that a static educational system does not serve society very effectively. The systems approach can help education to serve its purpose more effectively. Education is dynamic, not static; the systems approach is dynamic, not static.

Generally, the systems approach in education, or any enterprise for that matter, can be thought of as consisting of the following processes:

1. Define goals and objectives.
2. Delineate constraints and conditions.
3. Establish measures of effectiveness (standards).
4. Synthesize alternative solutions.

1. David E. Barbee, *A Systems Approach to Community College Education* (Princeton, N.J.: Auerbach, 1972) pp. 3–4.
2. Ibid., p. 8.
3. J. Alan Thomas, *The Productive School: A Systems Analysis Approach to Educational Administration* (New York: John Wiley, 1971), p. 3.

5. Establish costs for each alternative.
6. Select the best alternative and implement.
7. Follow through, feedback for improvement.

In this book, techniques useful for conceptualizing and implementing this generalized systems approach, as well as other generalized concepts associated with educational systems, will be discussed.

CYBERNETICS

Much of systems theory is founded on the cybernetics concept. This concept is a very important part of communication and control known as *feedback*, wherein a system is monitored and consequently modified to change the outcome in accordance with desired goals and objectives. *Cybernetics* denotes the correlation of communication and control. Its concern is for the components of a system and how they function together. The term, coined by Norbert Wiener from the Greek word *kybernetes* ("steersman"), is also the root word for "governor."[4]

In cybernetics, there are two forms of systems. They are called closed and open. In a *closed system*, each component can be completely identified and documented. Closed systems are often used by scientists and engineers in laboratory work. In contrast, an *open system* is one that may receive material or substance from its environment, and may provide material or substance to its environment. Since there are so many potential outside influences in an educational system, open systems are more realistic from the educator's point of view. Sometimes these influences are small and seem to be inconsequential, but every experienced educator knows that even a seemingly inconsequential event cannot be taken for granted. An open system influences, and is influenced by, elements external to the system. Chapter 4 considers the facets of the open system by using a real-life example that demonstrates the reaction of an open system to external influences.

COMPUTERS, INFORMATION, DECISION-MAKING, AND SYSTEMS

Many people confuse the computer with systems, assuming that one must use a computer if he uses systems management. Fortunately for those

4. Norbert Wiener, *Cybernetics; or, Control and Communication in the Animal and the Machine* (Cambridge: MIT Press, 1961), p. 16.

uninitiated into the marvels of electronic data-processing, it is completely possible and often desirable to forget the computer in the systems approach. Van Dusseldorp, Richardson, and Foley have made this point quite clear:

> It must be understood that systems conceptualizations, whether presented verbally, in printed form, pictorially, or in the form of computerized models, are only instruments designed to convey information for the facilitation of decision-making.[5]

Two important ideas, *information* and *decision-making*, are brought out in this statement. The manager of an educational enterprise is a decision-maker and must have appropriate information in order to function. Such an educational enterprise may be any unit, ranging from the smallest self-contained classroom with the teacher as manager up to the largest school district in its entirety. Indeed, the school district fits the definition of a system given earlier. It is a multiplicity of components (schools, teachers, students, classrooms, equipment, principals, board members, and the general public, to name a few), which work together for a common purpose. Is it necessary for the self-contained classroom, which is by definition a system, to have a computer? The answer is no, although many teachers would consider such a device a valuable classroom addition.

What is the role of electronic data-processing in educational systems? The computer is a subsystem which, if properly managed, can provide a wealth of information to subsystem managers throughout the whole system. The computer is a component in the overall system. This important point is discussed in greater detail in Chapter 3.

QUANTITATIVE ANALYSIS AND SYSTEMS

Just as some people confuse computers with systems, others have a similar problem concerning quantitative analysis. It is a fact that a certain amount of mathematics may be involved in the systems approach. Generally, mathematics is applied at the component level. The designer of a home heating system must be familiar with certain mathematical laws of physics before he can decide just which components are best for a given set of conditions. The homeowner, who is the user of the system, does not have to know how the components work in order to derive full benefit from the system. As a matter of fact, the homeowner needs to know just two things: the *input* and *output* requirements. In this case the input requirement is

5. Ralph A. Van Dusseldorp, Duane E. Richardson, and Walter J. Foley, *Educational Decision-making through Operations Research* (Boston: Allyn & Bacon, 1971), p. 4.

FIGURE 1.1. The "Black Box" concept

the type of energy used, and the output requirement is the range of temperatures which the heating system must produce. Engineers use what is called the "black box" to illustrate this fact (see Figure 1.1).

What goes on in the black box is not always of paramount concern to the user. As long as input and output requirements are met, the user can realize benefit and remain essentially ignorant of the internal operations of the system. The concept of the black box originated in the field of electrical engineering and has been used to explain systems such as radar.

In education, however, the inner workings of the system can be very important to the user. Consider the self-contained classroom as an example. The manager-teacher must know the input and output requirements. With knowledge of these requirements, the manager-teacher can then manipulate the components to maximize efficiency and effectiveness. Although this might not be very complicated in the quantitative sense, many educational-system components can be quite quantitative indeed. School bus schedules and routes, accounts and payrolls, and special teacher assignments are amenable to quantitative analysis. Chapter 2 considers a number of quantitative techniques in detail, some of which may be readily understood and put to use and others of which will require additional study. Whether the reader actually applies any of these techniques or not, he should make himself familiar with them. This will provide for a better understanding of the interrelationships of system elements.

SYSTEMS ORGANIZATION

Managers in business, industry, and government have long known that they must perform certain major functions. In essence, these are planning, organizing, and controlling. Some management theorists would add other functions, such as actuating, directing, and motivating. Such an in-depth theoretical approach is not needed in this direction of systems organization.

Managers in any enterprise must decide what is going to be done. This is *planning*. Goals are set, objectives determined, and a method of realizing or attaining these goals and objectives is decided upon. There are good and

bad plans resulting from these decision-making procedures. In any organization, good plans should be generated in order to maximize the possibility
of attaining the goals that have been set.

One way of enhancing good planning is to use the systems approach—
the seven-element process presented earlier. Basically, this is similiar to
what scientists and engineers have done for years; it is the scientific method,

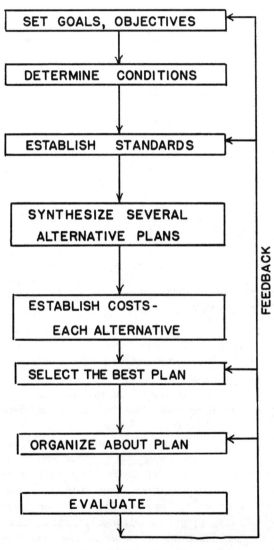

FIGURE 1.2. Planning model

as shown in Figure 1.2. It is important to understand that the best plan is not necessarily the least expensive one; it is the plan which realizes all the goals at the lowest cost.

Once a plan is determined, it is necessary to organize the enterprise for carrying out the plan. *Organizing* can be thought of in two sequential steps, structuring and allocating resources. *Structuring* is simply the so-called "chain-of-command" concept, which includes communication channels and responsibility assignments for each component of the system. The *allocation of resources* is the matching of personnel and materials to the structure. A word of warning is in order about resource allocation. All too often a person or resource is secured first, and then the structure is altered to accommodate the uniqueness of the person or resource. This can have disastrous results. For example, suppose a firm decides to put its accounting procedures on a computer base. Management rushes out and purchases a computer with all the peripheral equipment to handle the operation. Did they really buy the best resource to handle the job? Probably not. A systems-oriented firm would have first established goals and objectives for the computer-based system. It would have then structured the accounting organizational unit to meet the plan and selected the correct resource (computer) for its needs. Following such a procedure will minimize the chances of overbuying or underbuying. If the computer is to be used for other purposes, this will have to be considered in the planning process.

After the plan is made and the enterprise is organized to carry out the plan, the *control* function becomes important. The control function monitors the plan; that is, the plan is the blueprint for control. Included within this function are supervision, direction, motivation, and evaluation. As evaluation takes place, information is gained about the system. Whatever is learned from this process is sent back into other components of the system so that necessary changes can be made in order to carry out the plan effectively and efficiently.

The systems approach to educational management allows continuous focus on the goals and objectives as they are being considered throughout the management functions. If a deviation from the plan occurs in any function, resulting deviations may occur in every component of the system. Through systems-analysis techniques, the problem areas can be discovered and corrective action taken without unnecessary and costly delay. Models are often useful for this purpose, since a paper simulation can be generated and experiments performed at low cost. This technique allows managers to predict possible deviation points in advance, preparing them to deal with imbalance in the real-life implementation of the plan. Simulation techniques are discussed in Chapter 2.

MANAGEMENT AND ADMINISTRATION

For the purposes of this book, the word *manager* will have the same meaning as *administrator*. In the past, school officials have shunned the management label. In reality, many school administrators are, in fact, managers. They organize and control components of the educational system; they are in the line of the organization and are responsible for decision-making. This can be contrasted to staff personnel, who perform fact-finding and advisory functions. In this regard, a principal is most certainly the manager of his school.

SUMMARY

John McNamara has said that the "modern school administrator cannot afford to be uninformed concerning the operations of systems analysis."[6] He further noted that:

... systems theory has no value bias and is concerned only with achieving efficiently and effectively whatever it is that the system is designed to produce. ... if the goals and objectives of the system are directed toward humanistic needs, all of the strategies and controlling devices will be directed toward that purpose.[7]

Several important aspects of the systems approach have been introduced in this chapter. These considerations were clearly summarized by the American Association of School Administrators in the following manner:

1. The systems approach is goals oriented. The long-range and near-term goals of the educational enterprise are developed and continually reviewed for appropriateness throughout the process. These goals are kept in mind at all times.
2. There is an emphasis on planning in the systems approach. The system is dedicated to goal achievement, and careful planning is the key to success. Plans may be reviewed and revised as necessary in order to meet the goals.
3. Change is normal in the systems approach. Education is a dynamic process operating in a dynamic environment. When change is needed, it can be accommodated in the systems approach.
4. In the systems approach, the organization is the change mechanism. A static organization will not allow the systems approach to work effectively. In order to meet the overall goals of the organization, areas of responsibility may change as the plan is implemented.

6. John McNamara, *Systems Analysis for Effective School Administration* (New York: Parker, 1971), p. 5.
7. Ibid., p. 22.

5. The systems approach is model oriented. In order to keep costs at a realistic level, experiments are performed on paper, planning is done by use of models, and potential problem areas are identified with models.

6. In the systems approach a key concept is the identification of many alternatives. Ideally, all possible alternatives are determined and the best one is selected.

7. The systems approach is interdisciplinary. Empire-building is discouraged, for the system is no better than the components which must all work together for a common purpose.

8. Quantitative techniques are useful in the systems approach. With the emphasis on goal setting and planning, quantitative techniques are time and cost reduction agents, allowing a clear picture of the process for all components.

9. The systems approach is a rational decision-making process. In order to realize goals and their corresponding objectives, many decisions must be made. The systems approach allows a continuous and clear correlation of goals, objectives, and decisions which lead to rational outcomes.[8]

There is no difference, in effect, among the terms *systems approach*, *systems analysis*, *systems management*, and *systems theory*. In any new art or science, a multitude of terms arise, many with the same or similar meanings.

READING LIST

American Association of School Administrators. *Management by Objectives and Results*. Arlington, Va., 1973.

BARBEE, DAVID E. *A Systems Approach to Community College Education*. Princeton, N.J.: Auerbach, 1972.

MCNAMARA, JOHN. *Systems Analysis for Effective School Administration*. West Nyack, N.Y.: Parker, 1971.

THOMAS, J. ALLEN. *The Productive School: A Systems Analysis Approach to Educational Administration*. New York: John Wiley, 1971.

VAN DUSSELDORP, RALPH A.; RICHARDSON, DUANE E.; and FOLEY, WALTER J. *Educational Decision-making through Operations Research*. Boston: Allyn & Bacon, 1971.

WIENER, NORBERT. *Cybernetics; or, Control and Communication in the Animal and the Machine*. Cambridge: MIT Press, 1961.

———. *The Human Use of Human Beings*. New York: Doubleday Anchor Books, 1956.

8. *Management by Objectives and Results* (Arlington, Va.: American Association of School Administrators, 1973), p. 25.

CHAPTER 2

Techniques of Systems Analysis

There is a wealth of information in the literature about the many techniques of systems analysis. A comprehensive review and description of all these techniques would certainly be beyond the purpose for which this book was written.

The systems techniques included in this chapter were carefully selected on the basis of possessing, in the opinion of the authors, the greatest immediate promise for application to educational endeavors. Moreover, it was a further intention to give the reader an overview of a cross-section of the many techniques of systems analysis.

Included in the discussion of this chapter is a brief overview of a model-building technique called LOGOS (Language for Optimizing Graphically Ordered Systems). This technique has broad application to the total educational enterprise. It can be most appropriately applied by the curriculum-builder in designing comprehensive educational programs. Certainly it warrants serious consideration by administrators in the design of implementation procedures and strategies for the function of administration.

Critical Path Method (CPM) and Program Evaluation and Review Technique (PERT) will be outlined only in brief detail because the educational literature does give much attention to these two concepts. Despite the attention given to CPM and PERT, educators have been reluctant to use such techniques. Some have suggested that CPM and PERT are too difficult to use with educational projects and oftentimes involve a disproportionate amount of time. Contrary to this position, the authors believe that this opinion has resulted from a lack of information and understanding of the two concepts. In lieu of any further challenge of this position, a discussion of Precedence Diagramming Method (PDM) is presented in detail. Experience has shown that PDM is much simpler than CPM and PERT, and more appropriate for educational projects.

Like CPM and PERT, PDM is a monitoring system for the implementation of educational programs or projects. It facilitates decision-making and provides information for the most effective use of time, material and personnel.

Because of the increasing attention educators are giving to scientific decision-making, the discussion of flow charting, the Gantt Chart, and linear programming should provide a useful backdrop for future study. Finally, simulation is discussed in reference to many possible applications. Current uses of simulation show great promise for assisting in the solution of many of today's educational dilemmas.

LANGUAGE FOR OPTIMIZING GRAPHICALLY ORDERED SYSTEMS (LOGOS)

A unique systems language for flow-chart modeling is LOGOS, an acronym for Language for Optimizing Graphically Ordered Systems. LOGOS is a unique combination of alpha and numeric characters and geometric figures which communicate a systematic thought process. It is most appropriately used in model-building. "The thought expressed by a LOGOS flowchart is a conceptualization in the form of a graphic analog representing a real-life situation." [1]

In an effort to illustrate the symbols and methodology of the system, an application of LOGOS should provide the reader with an understanding of its mechanics and uses. The LOGOS model illustrated in Figure 2.1 will be employed for the purpose of this discussion. This model graphically depicts a comprehensive planning, programming, budgeting, and evaluation system which could be adapted to a public school system.

Upon careful examination of Figure 2.1, the reader will notice that the model employs a series of rectangles, some of which are contained in even larger rectangles. Moreover, the larger rectangles are circumscribed by even larger rectangles. All of these rectangles, in combination with a narrative description and a point-numeric code, are called *functions*. Functions can also be called subsubsystems, subsystems, systems, supersystems, suprasystems, and metasystems. The appropriateness of any one of these labels will depend upon a function's location in the total system. [2]

To illustrate the various functional descriptions, the reader is referred to the function "Describe Material Needs," which is identified by the point-numeric code 2.6.1 of Figure 2.1. This function could be called a *system*. The larger function 2.6 would then be called a *supersystem*; 2.0 would be identified as a *suprasystem*; and finally, if this total model were contained

1. Leonard C. Silvern, "LOGOS: A System for Flowchart Modeling," *Educational Technology* 9, No. 6 (June 1969): 18.

2. Ibid., pp. 18–19.

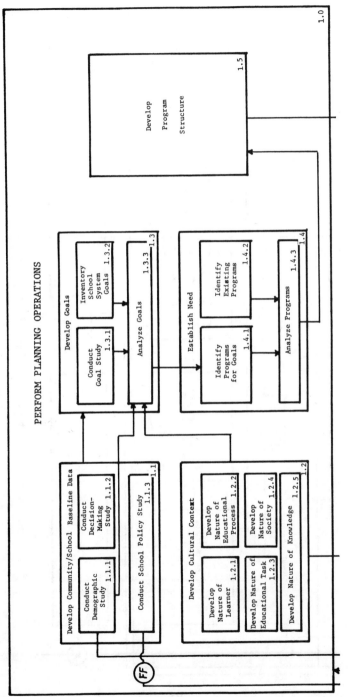

FIGURE 2.1. LOGOS model

23

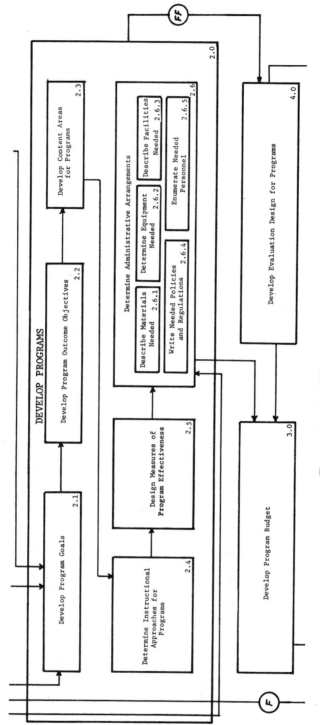

FIGURE 2.1. LOGOS model (continued)

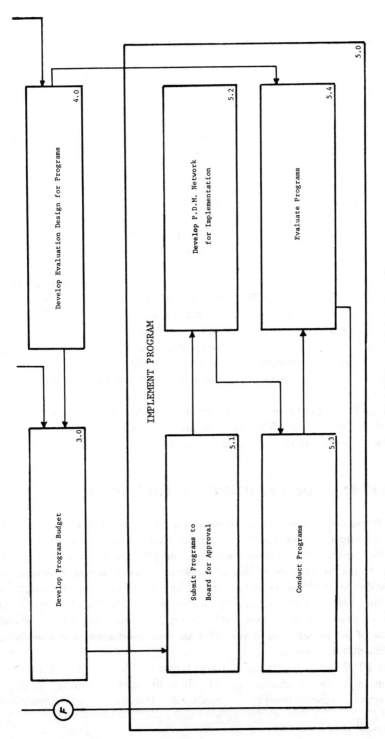

FIGURE 2.1. LOGOS model (continued)

in a larger rectangle as a function of a larger system, the larger system would be called a *metasystem* and coded 0.0. In general, "when one function is embedded in a larger function, the larger is a system and the smaller is a subsystem. If a subsystem consists of even smaller parts, these are also subsystems."[3] By observation one can see that through synthesis, subsystems 1.4.1, 1.4.2, and 1.4.3 of Figure 2.1 are a part of the larger system 1.4.

Data and information flow is represented in the system by arrows. The system will permit information to flow in only one direction. A function may either receive information (input) from another function or provide information (output) to another. Figure 2.1 shows the system 1.4.3 receiving inputs from systems 1.4.1 and 1.4.2 and providing an output of information to 1.5.

If an arrow is illustrating an input to a supersystem, this input is provided to all systems of the supersystem. An example can be seen in Figure 2.1. Notice that the output of 2.5 is also an input of 2.6. Therefore, it is an input to 2.6.1, 2.6.2, 2.6.3, 2.6.4, and 2.6.5.

Another symbol used is the circle. When arrows are extremely long, or if an arrow is to be continued on a second page, a circle with a point-numeric code included may be used to show an output to, or an input from, another function.

This brief discussion of LOGOS should provide the reader with a general understanding of the concept and its possible uses. An extended application of this modeling technique is discussed in Chapter 3.

PRECEDENCE DIAGRAMMING METHOD (PDM)

Precedence Diagramming Method is a useful methodology for scheduling, monitoring, and controlling educational projects or a sequence of activities. Specifically, it is used in the initial planning phases of an educational project for determining the interrelationships of all project activities; it provides for continuous monitoring of project activities throughout the life of the project where resource allocations and reallocation decisions must be readily made; it serves as a methodology for controlling the inputs and outputs of a project; and it provides an easy mechanism for budgetary planning and monitoring.

In PDM, all activities of a project can be illustrated so that their relationships, one to another, are clearly delineated. Moreover, the logic of all activity dependencies can be analyzed. "If a project can be described

3. Ibid., p. 19.

as a group of interrelated activities and if reasonable time durations can be assigned to each activity, it is susceptible to analysis by PDM."[4]

Network Planning and Logic

For the purposes of illustration, let us assume that a department of mathematics decides to install a new program for remediating mathematical skill deficiencies of students. This example is an oversimplification, but it will serve for illustrative purposes.

PDM has been identified as a systems technique to be used with the mathematics program. At the initial meeting of the departmental faculty, the program is discussed, as well as the activities appropriate for its implementation. The outcome of the meeting is the identification of the following program activities:

1. Assignment of tasks.
2. Review of literature.
3. Determination of budgetary constraints.
4. Inventory of student achievement.
5. Determination of scheduling constraints.
6. Development of curriculum.
7. Development of materials.
8. Approval of program for installation.

With the program activities identified, the PDM network is developed and illustrated in Figure 2.2.

Notice that each activity of the project has been included in the network of Figure 2.2. The rectangular boxes containing the activity descriptions are called *descriptors.* At each end of the descriptors are small squares called *nodes,* which are used for activity identification. The activity description "Review Literature" can also be described by the unique node identification numbers 15–20.

All work of the project moves from left to right, and the relationship of one activity to another is illustrated through the use of arrows. Arrows show the movement of work and information in the network and serve as restraints on project activities; that is, activities 15–20, 25–30, 35–40, and 45–50 cannot begin until activity 5–10 is completed. Similarly, activity 55–60 cannot begin until all of the above activities have been completed.

Once the integrity of the network logic has been established, activity time allocations are assigned. These time assignments are usually based

4. Paul A. Montello, "PDM: A System for Educational Management," *Educational Technology* 9, No. 11 (December 1971): 62.

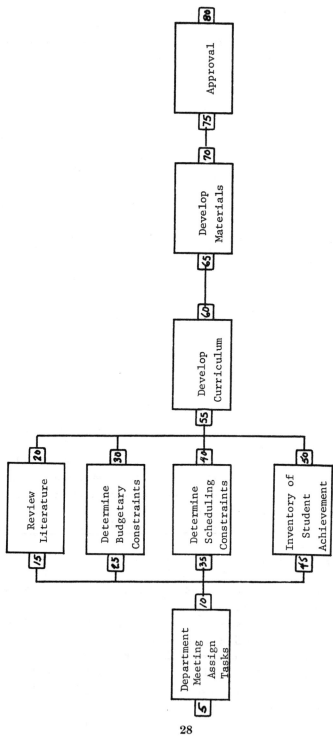

FIGURE 2.2. PDM network

upon good judgment and previous experience with similar activities. Time assignments can also be used as deadlines. Time assignments are placed below the descriptors, as illustrated in Figure 2.3.* Although any appropriate unit of time can be used, the unit chosen must be consistent throughout the network. The unit of time utilized in Figure 2.3 is day.

Usually, activity A begins on day 1 of the project. Figure 2.3 shows this notation above node 5. This notation identifies the earliest possible time (early start, ES) that activity A can begin. With a duration of two days for activity A, the earliest possible time that activity A can be completed (early finish, EF) is $1 + 2 = 3$, or the beginning of the third day, and is shown above node 10. The ES for activities B, C, D, and E must be 3 because they are restrained by A. The EF times for activities B, C, D, and E become $3 + 60 = 63$, $3 + 10 = 13$, $3 + 20 = 23$, and $3 + 15 = 18$, respectively. Activity F cannot begin until all the restraining activities of F have been completed. Therefore, the ES time of F must be 63. In a similar manner, the EF times for F, G, and H become $63 + 60 = 123$, $123 + 90 = 213$, and $213 + 10 = 223$, respectively.

Figure 2.4 illustrates the network with the inclusion of the late start (LS) and late finish (LF) times.

The EF time of 223 for H can also be called the latest possible time by which H must be completed. This is due to the fact that the last activity is always on the critical path. This logic will become self-evident when the critical path is discussed. Therefore, the LF time of H is 223 and is shown below node 80. It is now possible to work back through the network to compute all LS and LF times by employing the simple formula: $LS = LF - d$ where d is the activity duration. Thus, the LS time for H becomes $LS = 223 - 10$, or 213. The LS time for H is also the LF time for G; therefore, the LS time of G is 123. In a similar manner, the LS times of F, E, D, C, B, are 63, 48, 43, 53, and 3, respectively. Notice that activities B, C, D, and E are dependent upon activity A. Thus, the smallest LS time of these activities is the LF time for A. By observation one can see that the LF time for A is 3, and $3 - 2 = 1$ is the LS time for A.

Figure 2.5 illustrates the last value to be calculated to complete the network. This value appears under the duration time and is called total float (TF). "Total float is the time differential between the time available for an activity and the actual time required to complete the work."[5] The computation of TF for each activity can be easily obtained by determining the difference between the ES time of an activity and the LS time. As an example, the TF for activity E is $48 - 3$, or 45 days.

---

* To simplify the discussion, the activity descriptions in Figure 2.2 were replaced by capital letters in Figure 2.3.

5. Ibid., p. 63.

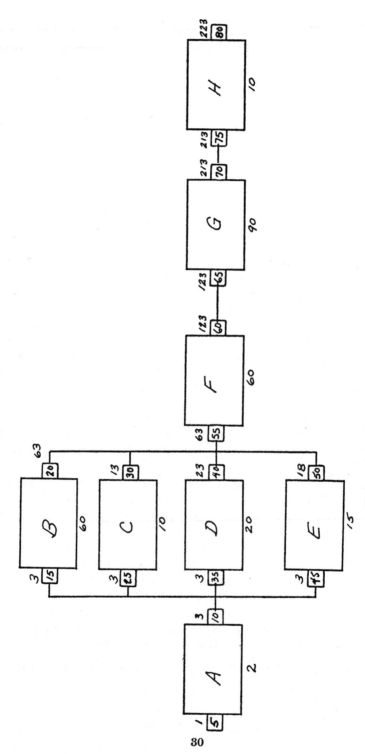

FIGURE 2.3. PDM network with time assignments

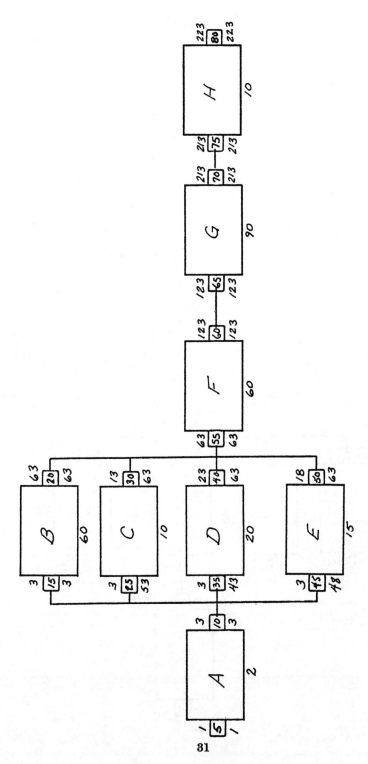

FIGURE 2.4. PDM network with late start and late finish times

31

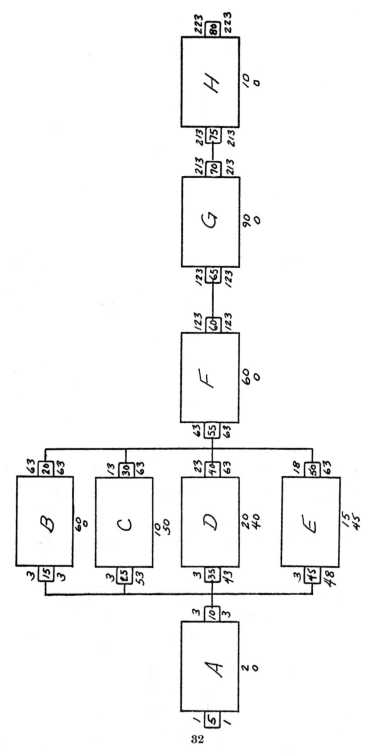

FIGURE 2.5. PDM network with total float

32

The critical path of the network is represented by the series of activities where the total float is 0. Activities on the critical path must begin on their ES times and must be completed by their LF times.

The critical path in the example can be identified by observations. It includes activities: A, B, F, G, and H.

PDM is, perhaps, the best systems planning and monitoring technique for educational projects. It provides for more effective utilization of scarce resources, including time, materials, and personnel. "It facilitates many decisions for re-allocation of resources; it holds projects to a time schedule; it aids in budgeting; and it provides immediate cost-time data for alternative activities of a project."[6]

CRITICAL PATH METHOD (CPM)

Critical Path Method is very similar to PDM, possessing nearly all of the same characteristics and attributes. The primary difference between the two techniques relates to the way in which activities are graphically represented. Dummy arrows are not used in PDM. These differences will be clarified in the discussion that follows.

CPM was developed by operations research personnel of the E. I. du Pont de Nemours Corporation and the Remington Rand Corporation. During 1956 Du Pont was engaged in planning the construction of major chemical plants and was concerned about the great amount of time it would take to realize output from the production of these plants. In an effort to shorten this time period, CPM was developed and successfully employed in the construction projects. Since this time, the construction industry and governments have adopted its use for a wide variety of projects. More recently it found its way into the management of educational projects and has been used successfully for such projects. A description of CPM can be best provided by relating its use to a sample educational project.

A local school system has decided to completely revamp its social-sciences program for junior and senior schools. After some thought, the curriculum director determines that the activities listed below are appropriate for completing the project. The duration time for each activity is given in weeks.

The curriculum director has the assistance of various faculty members and school social-science chairmen. Accordingly, he decides to divide the work among them. Also, he sees that things could be done in less than the sixty-week total. The sequence of activities, however, cannot be violated:

6. Ibid., p. 64.

	Activity	Time Estimate
a.	Decide general purposes of the new program.	2
b.	Determine goals for history (and government).	5
c.	Determine goals for sociology.	3
d.	Determine goals for economics.	3
e.	Plan history program for junior high schools.	4
f.	Plan history program for senior high schools.	5
g.	Plan sociology program for junior high schools.	5
h.	Plan sociology program for senior high schools.	3
i.	Plan economics program for junior high schools.	2
j.	Plan economics program for senior high schools.	5
k.	Coordinate the programs.	4
l.	Replan history program, select text and materials.	6
m.	Replan sociology.	5
n.	Replan economics.	5
o.	Implement program.	3
		60

no goals can be determined before general purposes are set, no program can be planned before goals are known, coordination must follow programs, replanning must follow coordination, and implementation comes last. This can be represented as follows:

Activity	Prerequisite	Time Estimate
a	—	2
b	a	5
c	a	3
d	a	3
e	b	4
f	b	5
g	c	5
h	c	3
i	d	2
j	d	5
k	e, f, g, h, i, j	4
l	k	6
m	k	5
n	k	5
o	l, m, n	3

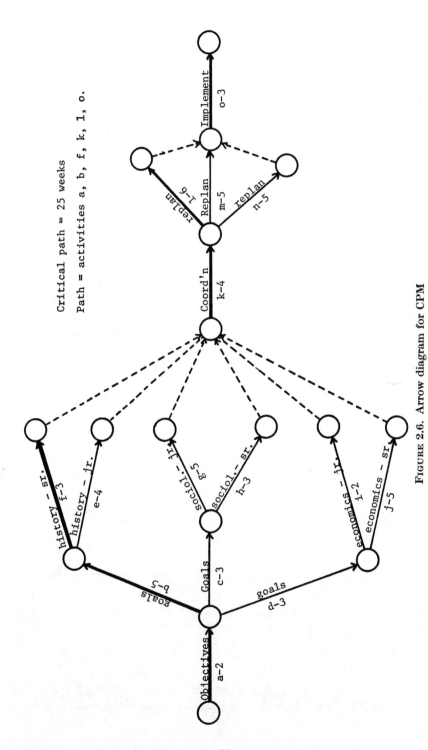

Critical path = 25 weeks

Path = activities a, b, f, k, 1, o.

FIGURE 2.6. Arrow diagram for CPM

Another way of illustrating the project is by an arrow diagram (see Figure 2.6. The bold set of arrows represents the critical path, or the longest possible path. Note that several activities are going on simultaneously. The diagram serves two purposes: first, it is a plan of how to undertake a project; second, it is a control device. As each activity is completed or partially completed, an appropriate symbol can be made on the diagram to indicate project status.

Listed below are the symbols used in CPM.

Symbol	Meaning
⟶ (solid arrow)	Job (activity)
○ (node)	Begin/end points
⇢ (dashed arrow)	Dummy (for connection only)

Dummies are particularly significant. The instance where two paths are parallel is generally better illustrated with a dummy in order to bound each arrow with a unique pair of nodes. The example shown in Figure 2.7 will help to clarify this point.

Examples showing the ES, EF, LS, LF, d, and F are illustrated below and in Figure 2.8. The critical path is identified by those activities where $T = 0$.

Activity	Predecessors	Duration
a	—	2
b	—	4
c	a, b	6
d	b	5

POOR   BETTER

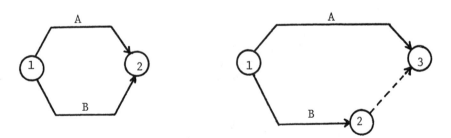

FIGURE 2.7. Method by which activities can be bounded by unique nodes

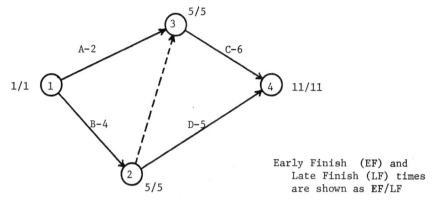

FIGURE 2.8. CPM diagram, showing the use of a dummy arrow. The critical path is B–C, with a duration of ten time units. Paths A–C and B–D are shorter.

The PDM technique discussed earlier can be used to convey the same information. In fact, the authors of this work prefer PDM over CPM, but they have included CPM because of its popularity. ES, LS, EF, and LF times are calculated in CPM as they are in PDM. The formula for float is: $F = LF - ES - d$.

PROGRAM EVALUATION REVIEW TECHNIQUE (PERT)

The theory basic to PERT is similar to PDM and CPM. The important difference is that computations in PERT are associated with statistical probabilities. By coincidence, both PERT and CPM were developed at approximately the same time. Radcliff, Kawal, and Stephenson have succinctly outlined the development of PERT.

At about the same time that DuPont was developing CPM, the Special Projects Office of the U.S. Navy was involved in a study of planning and control on the Polaris missile project. This project, called the Fleet Ballistic Missile Program, involved contract work by over 3000 individual companies. The Navy retained the consulting firm Booz, Allen, and Hamilton to work on this complex problem. In July of 1958, the Polaris team published a report of Phase I—Program Evaluation and Review Technique, which has become known as PERT. The original network planning concept and derivation of the mathematics for PERT were largely the work of Dr. C. E. Clark. This first report outlined the new system. In September of 1958, Phase II of PERT was published giving further details of the method, along with some modifications, and also described actual applications and installations of PERT. The success of PERT was startling; the Polaris missile project was completed almost two years ahead of the original schedule.[7]

In education, PERT has been given considerable attention; however, its use has been limited. This is, perhaps, due to the difficulty in making the mathematical calculation relative to time. In CPM, there is only one time estimate for each activity. There is an obvious weakness inherent in a single time estimate. One way of resolving this problem is to use multiple time estimates and work with an average. PERT does just this. In Figure 2.9, there is an arrow diagram with PERT time estimates. The smallest number is called the optimistic time; the largest number is called the pessimistic time; and the middle number is called the most likely time. These are often called A, B, and M times, respectively.

Average time (t_e) is calculated by using the formula:

$$t_e = \frac{A + 4M + B}{6}$$

Activity	Predecessor	Optimistic (A)	Most Likely (M)	Pessimistic (B)
A	—	5	7	8
B	A	4	6	8
C	A	3	4	8
D	B	3	5	9
E	B, C	2	6	7
F	D, E	3	6	7

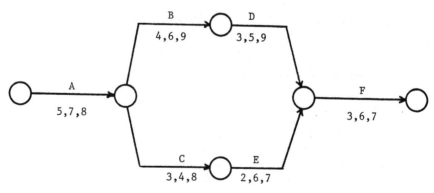

FIGURE 2.9. PERT diagram, with three time estimates per activity

7. Byron M. Radcliff, Donald E. Kawal, and Ralph J. Stephenson, *Critical Path Method* (Chicago: Cahners, 1967), pp. 7–8.

This is simply a weighted average with the most likely time (M) being given four times the weight of either A or B.

The main reason for using three time estimates is that a measure of variance or dispersion can be easily calculated. Variance (V_t) and the average (t_e) give much better information than a single time estimate as in CPM. The calculation of variance (V_t) can be derived by the following formula:

$$V_t = \frac{[(B - A)]^2}{3.2} \; *$$

Substituting the values of the first activity of Figure 2.9 into this formula, the variance is $[(8 - 5)/3.2]^2$, or about 0.88. Variances for the remaining activities are calculated in the same manner.

The critical path is determined by figuring the length of the longest paths, using t_e times. It is the path A–B–D–F. If we sum the variances along the critical path, we get the project variance, the square root of which will give the standard deviation for the total project.

Activity	t_e	V_t
A	6.83	0.88
B	6.00	1.56
D	5.33	3.52
F	5.67	1.56
	23.83	7.52

$$\text{Standard deviation} = \sqrt{V_t}$$
$$= \sqrt{7.52}$$
$$= \quad 2.74$$

Thus, we could expect to complete the project in 23.83 time units, with a standard deviation of 2.74 time units.

The reader will probably prefer to use PDM instead of PERT. A knowledge of the fundamental difference is important, however, since predetermined PERT plans are often available.

Figure 2.10 illustrates an actual example of a PERT network that was developed for conducting a research study. The analyses of the time data are presented in Tables 2.1 and 2.2. The critical path of the network is identified by those activities with a 1 in column 8 of Table 2.1. The probabilities for completing the project at a given time are illustrated in Table 2.2.

---

* The student of statistics should note that many users of PERT prefer 6.0 for a denominator. This results in a 0.003 level of significance for the total project standard-deviation calculation. Using 3.2 in the denominator will result in a 0.1 level of significance.

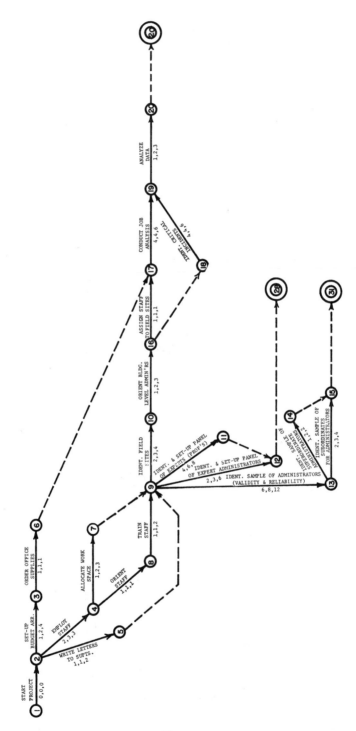

FIGURE 2.10. PERT network

40

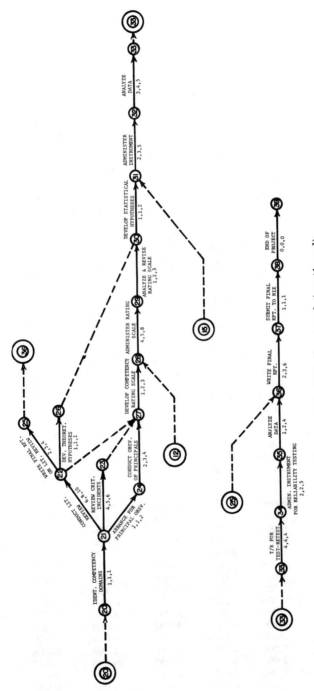

FIGURE 2.10. PERT network (continued)

41

TABLE 2.1

ANALYSES OF TIME DATA

Activity		Scheduled	Time	Durations	Mean	Critical	Critical
Begin	End	A	M	B	Duration	Variance	Paths
1	2	0.000	0.000	0.000	0.000	0.000	1
2	3	1.000	2.000	4.000	2.167	0.000	0
2	4	2.000	3.000	3.000	2.833	0.098	1
2	5	1.000	1.000	2.000	1.167	0.000	0
3	6	1.000	1.000	1.000	1.000	0.000	0
4	7	1.000	2.000	3.000	2.000	0.000	0
4	8	1.000	1.000	1.000	1.000	0.000	1
5	9	0.000	0.000	0.000	0.000	0.000	0
6	17	0.000	0.000	0.000	0.000	0.000	0
7	9	0.000	0.000	0.000	0.000	0.000	0
8	9	1.000	1.000	2.000	1.167	0.098	1
9	10	2.000	3.000	4.000	3.000	0.391	1
9	11	4.000	6.000	8.000	6.000	0.000	0
9	12	2.000	3.000	6.000	3.333	0.000	0
9	13	6.000	8.000	12.000	8.333	0.000	0
10	16	1.000	2.000	3.000	2.000	0.391	1
11	12	0.000	0.000	0.000	0.000	0.000	0
12	28	0.000	0.000	0.000	0.000	0.000	0
13	14	1.000	2.000	2.000	1.833	0.000	0
13	15	2.000	3.000	4.000	3.000	0.000	0
14	15	0.000	0.000	0.000	0.000	0.000	0
15	31	0.000	0.000	0.000	0.000	0.000	0
16	17	1.000	1.000	1.000	1.000	0.000	1
16	18	0.000	0.000	0.000	0.000	0.000	0
17	19	4.000	4.000	6.000	4.333	0.391	1
18	19	4.000	4.000	6.000	4.333	0.000	0
19	20	1.000	2.000	3.000	2.000	0.391	1
20	21	1.000	1.000	1.000	1.000	0.000	1
21	22	6.000	8.000	10.000	8.000	1.563	1
21	23	4.000	5.000	6.000	5.000	0.000	0
21	24	1.000	1.000	2.000	1.167	0.000	0
22	25	2.000	3.000	4.000	3.000	0.000	0
22	26	1.000	1.000	1.000	1.000	0.000	0
22	27	0.000	0.000	0.000	0.000	0.000	1
23	27	0.000	0.000	0.000	0.000	0.000	0
24	27	2.000	3.000	4.000	3.000	0.000	0
25	36	0.000	0.000	0.000	0.000	0.000	0
26	30	0.000	0.000	0.000	0.000	0.000	0
27	28	1.000	2.000	3.000	2.000	0.391	1
28	29	4.000	5.000	8.000	5.333	1.563	1
29	30	1.000	2.000	3.000	2.000	0.391	1
30	31	1.000	1.000	2.000	1.167	0.098	1
31	32	2.000	3.000	5.000	3.167	0.879	1
32	33	3.000	4.000	5.000	4.000	0.391	1
33	34	4.000	4.000	4.000	4.000	0.000	1
34	35	2.000	2.000	5.000	2.500	0.879	1
35	36	1.000	2.000	4.000	2.167	0.879	1
36	37	2.000	3.000	6.000	3.333	1.563	1
37	38	1.000	1.000	1.000	1.000	0.000	1
38	39	0.000	0.000	0.000	0.000	0.000	1

TABLE 2.2

PROBABILITIES FOR COMPLETING PROJECT AT GIVEN TIME

Time from Project Start	Probability of Completion
52.875 periods	10 percent
54.291 periods	20 percent
58.686 periods	30 percent
56.186 periods	40 percent
57.000 periods	50 percent
57.814 periods	60 percent
58.686 periods	70 percent
59.709 periods	80 percent
61.125 periods	90 percent
64.487 periods	99 percent

FLOW CHARTING

Flow charts are decision tools which are very useful in planning and control. Figure 2.11 shows the flow chart of a plan for teaching some particular topic.

In this flow chart there are only five symbols. In fact, most flow charts can be constructed using these basic five symbols. An oval is a terminal point, that is, a starting or stopping place, while input and output are symbolized by a parallelogram. Rectangles represent action, and diamond shapes represent decisions. Generally, these decisions are of the yes or no variety. Finally, the arrow shows the direction of flow.

It is possible to go back to some previously passed point by using the decision diamond and a reverse direction arrow. This is done in two instances in Figure 2.11. The principle involved is known as feedback (explained in Chapter 1), and it is this principle which makes the flow chart so useful to planners and controllers.

Computer analysts use flow charts for communication purposes. If one analyst prepares a chart, almost any other analyst, and most programmers, can read it and understand it in a very short time. The use of flow charts has been widely adopted in many fields, including that of educational planning, which has benefited from it.

Another form of flow chart is shown in Figure 2.12. This is called GERT, an acronym for Generalized Evaluation and Review Technique. GERT is a combination of CPM and the flow-charting technique discussed above.

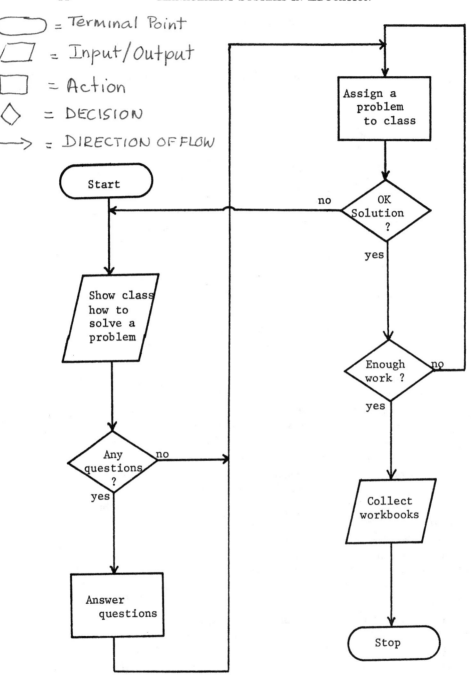

FIGURE 2.11. Example of flow chart

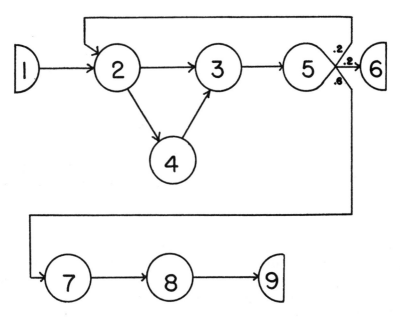

FIGURE 2.12. GERT diagram

The GERT symbols are similar to CPM or PDM. The semicircle, however, indicates a starting or stopping point. The circle represents a *stochastic* event, that is, one that is sure to happen, and goes on to the next (following) event. The third symbol, shown as event number 5 in Figure 2.12, is a decision point, or a *probabilistic* event. In the example illustrated in Figure 2.12, it has been determined that there is a 20 percent chance that the project will terminate, or stop, at event 6. There is a 20 percent chance that the project will have to be redone from event 2 onward, and there is a 60 percent chance that it will go straight through to completion. Probabilities are determined by polling people experienced in similar circumstances.*

The standard flow chart is probably best for use in education. If probability assignments are desired, it might be best to switch to a GERT-type chart. Either technique forces the user to: (1) consider every aspect of the situation; (2) write them down; and (3) demonstrate coordination with other events. A graphic plan can generally be quickly understood by everyone involved.

* A probability of 1.0 represents certainty, and a probability of 0.0 indicates that the event has no chance of taking place. An event with a 20 percent chance of occurring has a probability of 0.2; a 60 percent chance, 0.6.

GANTT CHARTS

Henry L. Gantt is generally credited with being the first to organize machine scheduling in factories. His objective was to eliminate long waiting times between machines in a production process, and his method, despite its simplicity, has been very effective. The Gantt Chart is not limited to machines and factories; it is useful in many instances for planning and controlling purposes.

The concept is simple. A vertical column, or list, is made of capacities, or necessary sources. Then a horizontal list is made showing time, generally in calendar days available. The work or task to be performed is then marked into the resulting chart. Figure 2.13 shows a sample project.

The "*V*" indicates current date. The number above each horizontal bar is a code to identify project activities. The heavy line indicates current progress; note that Sam is behind schedule by one day, and Ed is ahead by one. Wednesday would normally have been idle time for Sam, but since he is behind, the project manager has extended activity 112 by an additional day and indicated the extension by a crossed bar (block) which appear on Wednesday.

By carefully adjusting the physical dimensions of the chart, it is possible to show jobs taking less than one day (or week or month, according to the time unit decided upon). Although Gantt Charts are often used to schedule the valuable human resource, they are just as useful for scheduling

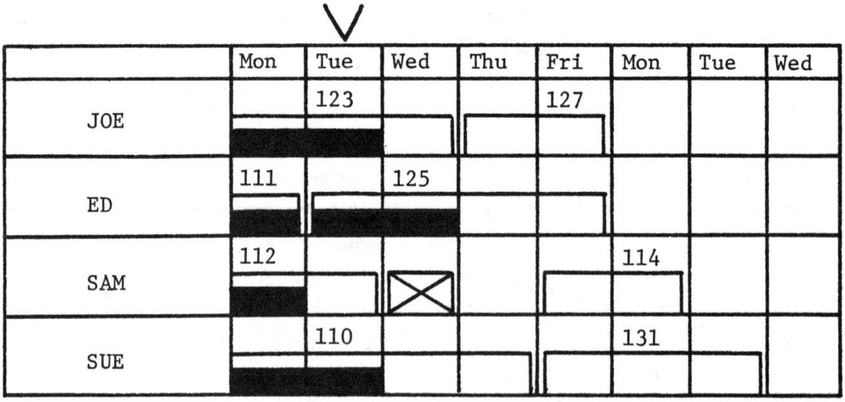

Key: Joe has been assigned jobs numbered 123 and 127. He is on schedule.

FIGURE 2.13. Sample Gantt chart

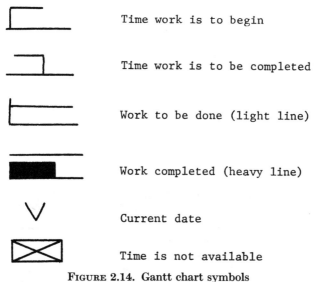

	Time work is to begin
	Time work is to be completed
	Work to be done (light line)
	Work completed (heavy line)
	Current date
	Time is not available

FIGURE 2.14. Gantt chart symbols

any other resource, especially machines and facilities. The charts are easily understood and give a graphic statement of plan and progress at a glance.

The basic symbols used in the Gantt Chart are shown in Figure 2.14. Users often modify these symbols to serve their own unique needs; and while this practice is to be commended, the users must remember to make up a "key" for use by other personnel.

LINEAR PROGRAMMING

Generally speaking, linear programming is useful in instances where the user wishes to optimize (either maximize or minimize) the use of certain limited resources under a given set of constraints. It is a rather sophisticated mathematical technique, often requiring the services of an expert in order to achieve a feasible solution. The reason for including the technique in this book is to present situations and problems for which linear programming can provide a solution, or a set of solutions. Hopefully, the reader will see the tremendous benefit to be derived from the use of the technique and pursue a more in-depth study of its application to educational problems. The following example is included for illustrative purposes.

The Centerville Consolidated School District has decided that a better racial balance among its elementary schools is in order. The board has determined that busing some students between schools will achieve the balance required. They are also concerned about conserving bus fuel. Therefore, they wish to minimize total miles traveled by bus for all students and maintain current total enrollment for each school at the same level. Furthermore, they want the same ratio of majority-to-minority students at all schools. Some information is needed in addition to the board's decisions; namely, a list of schools, a breakdown of students by race, and the mileage between schools. It is assumed that students to be bused will be allowed to report to the school closest to them each morning and be returned to that point in the afternoon. Table 2.3 shows that there are nine

TABLE 2.3

NUMBER OF STUDENTS AT EACH SET OF SCHOOLS

Minority Schools	Majority Schools
A—250 students	U—250 students
B—300	V—350
C—350	W—350
	X—250
	Y—300
	Z—300

hundred minority students at three schools, and eighteen hundred majority students at six schools. Ideally, there should be one hundred minority and two hundred majority students in each school, except for the practical fact that some schools would then be overcrowded while others would have too few students. To account for this, Table 2.4 illustrates the number of students who must be sent away from the minority schools to achieve the needed balance and the number of minority students that each majority school

TABLE 2.4

NUMBER OF STUDENTS TO BE BUSED

Number of Majority Students Sent to Minority Schools	Number of Minority Students Sent to Majority Schools
A—167 students	U—83 students
B—200	V—117
C—233	W—117
	X—83
	Y—100
	Z—100

MAJORITY SCHOOLS

	U	V	W	X	Y	Z
A	3	4	5	4	3	7
B	5	7	9	8	6	4
C	4	2	2	6	4	5

MINORITY SCHOOLS

FIGURE 2.15. Distances in miles between minority schools and majority schools

must accept. The balance is, of course, one-third minority and two-thirds majority at each school.

The problem has to be worked in two parts; one part to move minority students to majority schools, and the other part for the movement of majority students to formerly minority schools.

One more thing must be known. This is the distance from any one school to any other. Figure 2.15 shows the full set of distances in miles.

With the facts organized in a systematic manner, the problem can now be attacked by using a form of linear programming called the transportation method, which requires no higher mathematics. First, a movement schedule, called a tableau, is constructed (see Figure 2.16). This tableau is divided into cells. In the upper left corner of each cell is the mileage from school

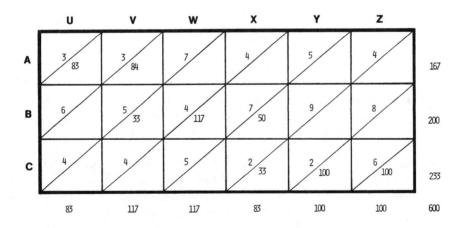

STUDENT-MILES: 3(83) + 3(84) + 5(33) + 4(117) + 7(50) + 2(33) + 2(100) + 6(100); OR 2,350

FIGURE 2.16. Transportation method of linear programming, tableau I

to school. Assignments of minority students to majority schools are made in accordance with the requirements in Table 2.4. In Figure 2.16, 83 students from school A have been assigned to school U (this represents the total number of students school U can accept); the remaining 84 students to be transferred from school A are assigned to school V; however, school V can still accept an additional 33 students for its total of 117; 33 students are assigned to school V from school B; the remaining 167 students of school B are assigned to schools W and X, 117 and 50 respectively; the 233 students from school C are assigned to schools X, Y, and Z. Total student-miles are then determined by multiplying the mileage of each "filled" cell by the number of students assigned to that cell, for every cell used. In tableau I, this works out to be 2,350 student-miles, or 3.92 miles per student, on the average. It would be desirable to reduce this average, if possible.

The object is to reallocate the students to cells which are not so distant, and some careful thought should be given to this matter. Observe that only eight cells are filled. Observe further that the cell A–Z shows four miles between the two schools, while the cell C–Z is six miles. If the hundred students transferring to school Z are taken from school A, significant mileage would be saved. In other words, students from C to Z would have traveled six hundred. If one hundred students were taken from school A, total student-miles would equal four hundred, for a two-hundred-mile saving. Making this maneuver causes too many students to be sent away from A. The tableau must be adjusted to account for this problem. The second adjustment will create the need for a third, and so on. After a while, however, a point will come when the tableau is balanced; and the new tableau thus created can be checked to see if any improvement has resulted. When no further improvements can be effected, the best solution has been found.

In Figure 2.17, tableau II shows the best solution possible. It does not take into consideration logistical problems, such as the size of vehicle required for each movement. Once this best solution has been reached, it may have to be modified slightly for logistical or other reasons.

The reader interested in pursuing this technique is encouraged to consult any introductory textbook on management science or operations research. There are several ways of telling when one has reached the best solution and several ways of arriving at the solution, and most texts will explain these methodologies. A knowledge of simple addition and multiplication, plus some good common sense, is all that is required to become proficient at such a technique.

In the example problem, the solution for moving the correct number of minority students to formerly majority schools has been found. Now it

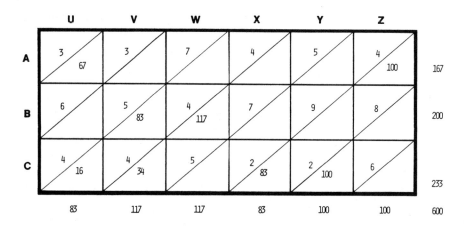

STUDENT-MILES FOR THIS TABLEAU: 2,050; OR 3.42 MILES PER STUDENT, ON THE AVERAGE.

FIGURE 2.17. Transportation method of linear programming, tableau II

is necessary to move majority students to the formerly minority schools by making tables and tableaus in a similar fashion. This information is then given to the school-system transportation director so that he can establish the actual schedules and routes. The example, of course, uses simple numbers and few schools. In reality, a problem such as this one would be more complicated, but the technique shown works well for any size.

Linear programming techniques can be used for many other purposes, such as teacher assignments, facility scheduling, and perhaps student scheduling. This is a powerful tool, and its full impact upon educational management has not yet been felt.

SIMULATION

Oftentimes in education, problems are so complex that any given solution may have catastrophic effects on a school system or systems. It is, therefore, necessary to test all the possible solutions through the use of simulation before any single one is actually implemented. One example would include the testing of several state funding procedures for the school sytems of a state. A solution which would provide the best funding model under certain conditions could be identified through the use of simulation.

Most often this type of simulation requires the use of a computer. In this regard, Martin defined computer simulation as "a logical-mathematical representation of a concept, system, or operation programmed for a solution on a high-speed electronic computer."[8] Moreover, according to Banghart:

Simulation is a technique that through modeling procedures attempts to develop an analogue of a set of inter-related activities. Simulation studies are usually conducted on computers. They permit the administrator to run an operation as a game before the actual activity is initiated. Many types of simulation studies have been conducted. The major effort has taken place in the military and business. The military has simulated war games dating back several centuries to Japanese and Prussian war experiences. The present military establishment conducts simulations of war games but also builds simulation trainers for such purposes as training jet pilots.[9]

Simulation shows great promise for application to educational situations. Its application to problems should be explored whenever (1) analytical tools are unavailable or inappropriate for finding solutions, (2) there is assurance that the system can be simulated, and (3) a large amount of computation is necessary.[10]

Fortunately, many problems in education that are susceptible to simulation are of such a nature that a computer is not necessary. For the purpose of illustrating the technique of simulation, one such problem will be used as an example.

The assistant superintendent for counseling services of the Midvale Unified School District wishes to know how many counselors to employ for the East Midvale High School, which is planned to open next fall.

Data from a similar school, already in operation, are available, and these are illustrated in Table 2.5. Interpreting the table, it can be seen that 3 percent of the counseling sessions last from one to five minutes, 4 percent last from six to ten minutes, and so forth. These data can be used to simulate the new high school after some minor conversions. First, the percentage figures must be rearranged in cumulative fashion. Then, a mean (average) time for each time category must be found. Finally, assignments of random numbers must be made. These changes are shown in Table 2.6.

The average time for each category of "Minutes per Session" is found by selecting the midpoint of the category; that is, the midpoint of the category of 1–5 minutes is 3 minutes.

Random number assignment is then determined by using the cumulative percentage figure for each category as the last random digit in the

8. F. F. Martin, *Computer Modeling and Simulation* (New York: John Wiley, 1968), p. 5.

9. Frank W. Banghart, *Educational Systems Analysis* (London: Macmillan, 1969), p. 4.

10. Martin, *Computer Modeling and Simulation*, pp. 15–16.

TABLE 2.5

REFERENCE DATA FOR LENGTH AND PER-
CENTAGE OF OCCURRENCE FOR COUNSELING
SESSIONS

Minutes per Counseling Session	Percentage of Occurrence
1–5	3
6–10	4
11–15	7
16–20	13
21–25	15
26–30	18
31–35	15
36–40	10
41–45	8
46–50	4
51–55	2
56–60	1
	100

TABLE 2.6

INITIAL RANDOM NUMBER ASSIGNMENT

Minutes per Session	Average Time	Cumulative Percentage*	Random Numbers
1–5	3	3	1–3
6–10	8	7	4–7
11–15	13	14	8–14
16–20	18	27	15–27
21–25	23	42	28–42
26–30	28	60	43–60
31–35	33	75	61–75
36–40	38	85	76–85
41–45	43	93	86–93
46–50	48	97	94–97
51–55	53	99	98–99
56–60	58	100	00

* Cumulative Percentages are derived from the accumulation of
percentages in the "Percentage of Occurrence" column of Table 2.5.

range of random numbers for each category. For example, 3 percent of the time it takes 1–5 minutes to serve a counselee; the random numbers 1, 2, and 3 represent this condition. Random numbers 4, 5, 6, and 7 represent the second category, which is 7 percent of the time, 8, 9, 10, 11, 12, 13, and 14 represent the third category, and so on.

Table 2.7 can now be used for the simulation, along with a table of random numbers. Such tables are found in many mathematics textbooks and reference books. The first three columns of Table 2.7 represent the first

TABLE 2.7
COUNSELING PROBLEM SIMULATION

Student	Random Number from Table	Time to Counsel	Second Random Number from Table
1	39	23	42
2	00	58	(6 students in first hour)
3	35	23	
4	04	8	
5	12	13	
6	11	13	
7	23	18	33
8	18	18	(4 students in second hour)
9	83	38	
10	35	23	
11	50	28	92
12	52	28	(10 students in third hour)
13	68	33	
14	29	23	
15	23	18	
16	40	23	
17	14	13	
18	96	48	
19	94	48	
20	54	28	
21	37	23	25
22	42	23	(4 students in fourth hour)
23	22	18	
24	28	23	
25	07	8	05
26	95	48	(2 students in fifth hour)
		668	

part of the simulation. Column 1 represents the students as they arrive for counseling. Column 2 contains the random numbers from a set of the first one hundred random numbers of a table.

Column 3, "Time to Counsel," is determined for each student by utilizing the random number assigned to each student in column 2 of Table 2.7 and by finding the average time for each student in Table 2.6 (column 2) appropriate for that random number. To illustrate, notice that student number 1 (Table 2.7) has a random number assignment of 39. Referring to column 4 of Table 2.6, you will find that 39 falls in the range of random numbers from 28 to 42. Therefore, the average time for student number 1 is twenty-three minutes. Similarly, 00 is the random number for student number 2, and from Table 2.6 (fourth column) it can be seen that this means the student will take fifty-eight minutes of counseling time. The rest of the third column of Table 2.7 is determined in a similar manner.

This demonstrates that the first twenty-six students to arrive will require a total of 668 minutes of counseling time, found by summing column 3 of Table 2.7. Although this information is useful, it is incomplete. The assistant superintendent knows more about the similar school—the average number of students who arrive for counseling in any given hour, plus the associated percentages. These data permit the development of Table 2.8.

TABLE 2.8

FINAL RANDOM NUMBER ASSIGNMENT

Average Arrivals per Hour	Percentage of Occurrence	Cumulative Percentage	Random Number Assignment
2	10	10	1–10
4	25	35	11–35
6	35	70	36–70
8	20	90	71–90
10	10	100	91–00

With the information in Table 2.8, column 4 of Table 2.7 can be constructed. From the table of random numbers, a new random number is selected. This number turned out to be 42. In the last column of Table 2.8, you will find the range in which 42 fits. It corresponds to an average arrival of six students. So, six students are associated by parentheses (column 4 of Table 2.7). The next random number from the table is selected. It is 33, which corresponds to an average of four students. The four are placed in parentheses. This procedure is continued to complete column 4 of Table 2.7.

The first and last columns of Table 2.8 and the third and fourth columns of Table 2.7 tell the assistant superintendent what he needs to know. Ten students will represent the greatest number that arrives in any one hour. Notice that column 4 of Table 2.8 shows a random number assignment of 91–00 for ten arrivals. Now referring to Table 2.7, the range 91–00 is associated with the random number 92 of column 4. This shows that ten students arrive during this hour for a total counseling time of 28 + 28 + 33 + 23 + 18 + 23 + 13 + 48 + 48 + 28, or 290 minutes. If the assistant superintendent wishes to avoid backlogs, he must have five counselors available to handle the situation, since a counselor can only work sixty minutes in an hour. If backlogs are acceptable, he might adjust the number. of counselors required accordingly.

Actually, the simulation should be run for about one thousand students instead of twenty-six. This would take only a few hours of desk time to compute and would be a reasonably accurate representation of what was going to happen at the new school. If better student counseling resulted the few hours would be well spent.

READING LIST

BANGHART, FRANK W. *Educational Systems Analysis*. London: Macmillan, 1969.

BOWMAN, E. H., and FELTER, R. B. *Analysis for Production and Operations Management*. Homewood, Ill.: Richard B. Irwin, 1967.

Educational Technology. June 1969 and June 1973 issues.

HUSSAIN, K. M., and HANDY, H. W. *Network Analysis for Educational Management*. Englewood Cliffs, N. J.: Prentice-Hall, 1969.

MARTIN, F. F. *Computer Modeling and Simulation*. New York: John Wiley, 1968.

MODER, J. J., and PHILLIPS, C. R. *Project Management with CPM and PERT*. New York: Van Nostrand Reinhold, 1970.

MONTELLO, PAUL A. "PDM: A System for Educational Management." *Educational Technology* 11, No. 12 (December 1971).

MONTELLO, PAUL A. "Systems Planning for Higher Education." In *Accountability: Systems Planning in Education*. (Creta D. Sabine, ed.) Homewood, Ill.: ETC Publications, 1973.

RADCLIFF, BYRON M.; BEAUMONT, M. J. S.; and LIVINGSTON, K. J. *Precedence Diagramming and Critical Path Method*. Lincoln, Nebr.: Industrial Research and Information Services Division, Nebraska Department of Economic Development, July 1969.

RADCLIFF, BYRON M.; KAWAL, DONALD E.; and STEPHENSON, RALPH J. *Critical Path Method*. Chicago: Cahners Publishing Co., 1967.

VAN DUSSELDORP, R. A.; RICHARDSON, D. E.; and FOLEY, W. J. *Educational Decision-making through Operations Research*. Boston: Allyn & Bacon, 1971.

Comprehensive Management Systems

There exist a number of comprehensive management systems that can appropriately be applied to the educational enterprise. Of this number, several have proven to be extremely useful in managing educational projects and systems. Specifically, Planning, Programming, Budgeting Systems (PPBS) have been adopted by hundreds of school systems as a total management system; Management Information Systems (MIS) are getting widespread attention from administrators, and have become a necessity in large city school systems; and Management by Objectives (MBO) is gaining acceptance in education as a comprehensive management system. This interest in MBO has resulted from the "Education for Results" movement, which is permeating all of education today.

PLANNING-PROGRAMMING-BUDGETING SYSTEM (PPBS)

Planning, Programming, Budgeting System has emerged out of a need to improve decision-making relative to individual corporate and institutional product priorities, allocations of resources, and alternative courses of action in implementing priority decisions. Organizations like multinational corporations, federal and state governments, and public educational institutions are characterized by a vast diversity of activities, all of which are directed toward the overall institutional or organizational purposes. The implementation of these activities requires many and varied resources. The personnel of many such institutions number into the tens of thousands while budgets total in the billions of dollars. Certainly, these institutions need to employ techniques of systematic planning so that their multitudinous activities can be directed toward a common and agreed-upon goal.

57

If this is done, scarce resources of time, material, and personnel can be efficiently and effectively utilized.

At a time when the use of public resources for education is coming under the close scrutiny of Congress, state legislatures, and the public, it is incumbent upon educators to draw from technology the best methodologies available for making programs relevant and for the maximum utilization of available resources. PPBS is such a methodology.

Although the roots of PPBS can be traced back to the War Production Board of 1942 and to private industry even before that time,[1] the concept emerged most noticeably as the technology of the Department of Defense during the McNamara era of the early and mid-1960s.[2] The impetus for PPBS in education can be traced to the efforts of members of the Association of School Business Officials (ASBO). This educational interest in PPBS occurred in the late 1960s at a time when the purse strings of expanded educational opportunity were tightly bound. The three-year $680,000 federally funded project of ASBO was initiated for the purpose of developing "a conceptual design for an integrated system of planning-programming-budgeting and evaluating (PPBES) which would be appropriate for local school systems."[3]

Many educational writers have defined PPBS. According to McGivney and Hedges, "PPBS is a methodology for improving decisions that have to do with the allocation of scarce resources to attain maximum satisfaction of our unlimited wants."[4] Moreover, Alioto and Jungherr state, "The Planning-Programming-Budgeting System is a framework for planning—a way of organizing information and analysis in a systematic fashion so that the consequences of particular choices can be seen as clearly as possible."[5] Farmer was more explicit when he said that PPBS is a system for:

Planning—the selection or identification of the overall long-range objectives of the organization and the systematic analysis of various courses of action in terms of relative costs and benefits.
Programming—deciding on the specific course of action to be followed in carrying out planning decisions.

1. David Novick, *Origin and History of Program Budgeting* (Santa Monica, Calif.: Rand Corp., 1966), p. 3.

2. Harry J. Hartley, *Educational Planning-Programming-Budgeting: A Systems Approach* (Englewood Cliffs, N.J.: Prentice-Hall, 1968), p. 78.

3. William H. Curtis, *Educational Resources Management System* (Chicago: Research Corporation of the Association of School Business Officials, 1971), p. 9.

4. Joseph H. McGivney and Robert E. Hedges, *An Introduction to PPBS* (Columbus: Charles E. Merrill, 1972), p. 7.

5. Robert F. Alioto and J. A. Jungherr, *Operational PPBS for Education* (New York: Harper & Row, 1971), p. 9.

Budgeting—translating planning and programming decisions into specific financial plans.[6]

In a short treatment of PPBS, Kaufman observed:

Most discussions of PPBS note that as a tool it is best used for taking the objectives of education, identifying alternate courses of action intended to meet the objectives, . . . and ranking the various alternate choices . . . in terms of their respective cost and benefits; then choices among the alternatives may be made on a more rational and empirical basis, and it is possible to derive a budget based on cost of achieving objectives.[7]

Generally, these definitions are very broad and tend to separate planning as a distinct activity apart from programming or budgeting. This appears to imply that there is an ordering in the process whereby planning activities must be finalized before programming can be initiated, and budgeting follows programming. Experience has shown, at least in the pilot school systems of the ASBO project, that the implementation of the methodology is a totally integrative process involving all levels and areas of expertise of a school system. This experience has justified the changing of the concept from PPBS to Educational Resources Management System (ERMS) by the research group at ASBO.

For the purposes of this discussion of PPBS, the concept is defined as:

A totally integrative process of comprehensive curriculum development where alternative courses of action are developed for identified goals. The courses of action which provide for the maximization of effectiveness with the minimization of resources are selected for implementation.

This definition does not imply that PPB is simply a cost-accounting system wherein efficiency is paramount. On the contrary, PPB is a system that gives focus to the purposes of the school or school system. It makes goals come alive in the process of curriculum development. Simply stated, it is a system for aiding the decision-making processes for activities of schools. It is not humanizing, nor is it dehumanizing. It is neutral. It simply aids the decision-makers in organizing information for making better decisions.

6. James Farmer, *Why Planning, Programming, Budgeting Systems in Higher Education?* (Boulder, Colo.: Western Interstate Commission for Higher Education, February 1970), p. 7.

7. Robert A. Kaufman, *Educational System Planning* (Englewood Cliffs, N.J.: Prentice-Hall, 1972), p. 132.

Although there is no single best approach to implementing a PPB system, there are more primary concepts that will be explored in this discussion. Basic to the discussion of these concepts is a set of assumptions, which were aptly stated by the research group of ASBO. These assumptions include:

1. The resources available to a school district are less than equal to the demands of that district.
2. The school district exists to produce a set of outcomes—to achieve certain objectives expressed as specific changes in characteristics of the learners.
3. Objectives of a school district can be achieved theoretically in a multitude of ways (program plans), some of which are more effective than others.
4. Productivity of a school district can be increased by the organization of learning activities and supporting services into programs specifically directed toward achieving previously defined goals and objectives.
5. Better decisions regarding the selection of program plans and greater benefits from their operation result when the costs thereof are considered on a long-term (multi-year) basis.
6. Better decisions regarding the selection of program plans and greater benefits from their application result when outcomes are related methodically to objectives.[8]

The logic of the discussion to follow is based upon these assumptions.

It is generally agreed that no generalized PPBS model will fit all school systems. Moreover, many writers suggest that an individual model unique to each system be developed. Unfortunately, this important consideration forces many researchers and authors to write in general terms and at a theoretical level beyond the understanding of the beginning student of PPBS. For these reasons, the following discussion will present an individual model of PPBS in the hope that the concepts discussed will provide for better understanding of how the system and ideas apply to the real world. The model that will be discussed is illustrated as a LOGOS model in Figure 3.1. The parenthetical numbers in the passages that follow refer to elements in the figure.

Perform Planning Operations (1.0)

The concept of planning connotes a process that has something to do with future events. In this regard, Dror defined *planning* as "the process of preparing a set of decisions for action in the future, directed at achieving

8. Curtis, *Educational Resources Management System*, pp. 37–39.

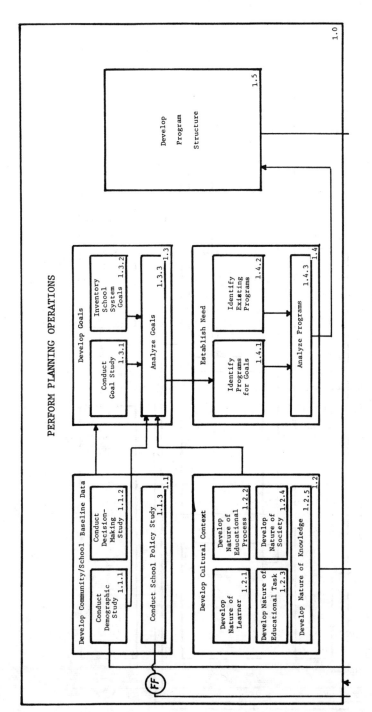

FIGURE 3.1. PPBS model

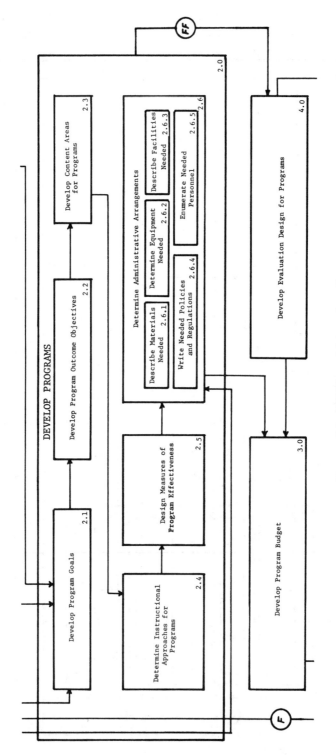

FIGURE 3.1. PPBS model (continued)

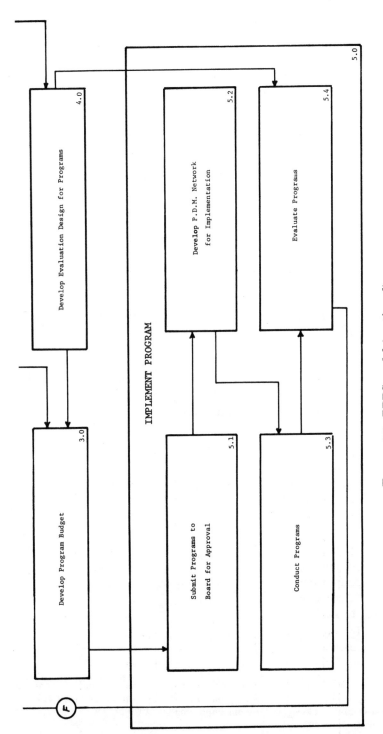

FIGURE 3.1. PPBS model (continued)

goals by optimal means."[9] He further stated that it is a process wherein continuous activity takes place for preparing a set of decisions to be approved and implemented by other units of the organization. With regard to a set of decisions, he said: "While planning is a kind of decision-making, its specific characteristic in this respect is its dealing with a set of decisions, i.e., a matrix of interdependent and sequential series of systematically related decisions."[10] "Decisions for action" implies that the planning process is "action" or "execution" oriented. The actions are designed for the future and are directed toward goal realization through the most optimal means.[11]

Develop Community/School Baseline Data (1.1)

The reader is reminded that a PPB system in education is a comprehensive curriculum-development system. Basic to curriculum, and part of the planning process, is the identification of important information about the school system and community. Included in subsystem 1.1 of Figure 3.1 are three subsubsystems which provide this information.

A complete understanding of the school-community can be gained from a comprehensive demographic study (1.1.1). This study should include efforts to identify each of the following:

1. Population (Current and Projected)
   (a) age groups
   (b) ethnic groups
   (c) racial groups
   (d) family units
   (e) educational level
   (f) birth rate
   (g) in-migration
   (h) out-migration
2. Employment (Current and Projected)
   (a) unskilled
   (b) skilled
   (c) professional
   (d) male
   (e) female

9. Yehezkel Dror, "The Planning Process: A Facet Design," in *Planning, Programming, Budgeting: A Systems Approach to Management*, ed. Fremont J. Lyden and Ernest G. Miller (Chicago: Markham Publishing Co., 1970), p. 99.
10. Ibid., p. 99.
11. Ibid., p. 100.

3. Civic and Religious
   (a) civic groups
   (b) church organizations
4. Geographic and Housing (Current and Projected)
   (a) percentage of rural houses
   (b) percentage of urban houses
   (c) number of single-family dwellings
   (d) number of multiple-family dwellings
   (e) number of mobile homes
5. Incomes (Current and Projected)
   (a) per capita income
   (b) family income
6. Business and Industry
   (a) number and types of businesses
   (b) types of industry

Output from the demographic study data for two important considerations. First of all, these data must be taken into account in all phases of program development (2.0). Additionally, the demographic study will identify many of the important constituencies, or groups, of the school-community. Members of these constituencies will then be included in the sample of participants of the goal study (1.3.1).

Other important baseline data include the identification of community people who are instrumental in all significant community decisions. To determine the identity of these persons, a decisions study is conducted (1.1.2). Oftentimes, these individuals are referred to as the "power structure" of the community. Their identity is important for the goal study. Certainly, any effort to set future direction for a school system should include the community decision-makers. The involvement of these persons is a must if any programmatic changes are anticipated. Without their support, it is unlikely that important changes will be accepted by the community.

In anticipation of program modification or introduction of new programs, it is necessary to conduct a school-board policy review and analysis (1.1.3). This analysis would identify possible policy conflicts and deficiencies. As an example, a high-priority goal of the goal study might be to provide a comprehensive drug-education program. The analysis of school-board policies may reveal an absence of policy on drug abuse or drug education. It therefore becomes necessary to develop appropriate policies so that the administration of the school system can implement drug-education programs in conformity with the directions set by the school board.

Develop Cultural Context (1.2)

The cultural context is a philosophical position or system of values from which the goals and programs of the school system are influenced. The development of the cultural context is based upon an interpretation of the nature of the learner, the nature of the existing society and the society to come, the nature of the learning process, the nature of knowledge, and the educational task. An example of an interpretation of the cultural context might be:

The interpretation of the cultural context focuses on the products of past experiences. These products suggest that knowledge is the outcome of rigorous inquiry which originates within the framework of human experiences. Knowledge is tentative and must be continuously tested and re-examined. Human experiences for every individual are regarded as unique. Moreover, the individual is unique in needs, interests, and abilities. Hence, society is in an ever changing environment. The mission of the . . . [school system], therefore, becomes the task of creating or finding existing conditions in which an individual may gain insight and understanding of the variables affecting social progress. Here social progress is defined as changes which enable human beings to lead the kinds of lives which promote human excellence. It is further defined as the changes in institutional relationships which free individuals and social groups from arbitrary restrictions in the free exchange and use of ideas. With these interpretations of certain elements of the cultural context as a basis, the learning process becomes the gaining of insight and personal understanding of the variables affecting social progress.[12]

Develop Goals (1.3)

Subsystem 1.2 utilizes the input data and information from subsystem 1.1. These inputs are needed, as suggested previously, for identifying a possible sample for the goals survey (1.3.1). One of the best techniques for conducting the goal survey is the Delphi methodology. This method will insure a high degree of consensus of the many constituencies of the school-community relative to the direction that the school system should take. Moreover, this method does not require that the participants be brought into a face-to-face confrontation to reach such a consensus.

Concurrently, a survey of existing system goals (1.3.2) is conducted in an effort to identify directions previously set by the school board. The finalization of systemwide goals is made by conducting an analysis of newly

12. Paul A. Montello, "Systems Planning for Higher Education," in *Accountability: Systems Planning in Education*, ed. Creta D. Sabine (Homewood, Ill.: ETC Publications, 1973), p. 125.

established goals and existing goals (1.3.3). Here it is necessary for the board of education to decide which goals it wishes to draw from the goal study and which it desires to retain from previous goal statements. This analysis cannot be conducted without a thorough consideration of the demographic information and an interpretation of the cultural context. The goal statements then become the basis for establishing program needs.

Establish Need (1.4)

Just as the basic unit of our monetary system is the dollar, the basic unit of need is program. Need is defined in terms of programs, and it is represented by the difference between programs appropriate for newly established goals (1.4.1) and existing programs (1.4.2). Therefore, the task of establishing need is to identify programs appropriate for newly established goals, enumerate existing programs, and analyze the two sets of programs (1.4.3) to determine which should be added and which should be retained, modified, or discarded.

Develop Program Structure (1.5)

The relationship of major program areas and subprograms can be illustrated in a program structure. The program structure represents the overall conceptual organization of the school system. This representation clearly delineates school-system activities in relation to goals. Figure 3.2 illustrates a portion of a program structure. Generally, the program structure is a hierarchical configuration, which provides a framework for individual program development, budgeting, and evaluation.

Develop Programs (2.0)

Programming, like planning, is a process. It is a process for developing program plans that can be utilized to move an institution (school or school system) toward the realization of its goals. Moreover, "programming is the link between planning and the decision to authorize resources to implement programs." [13] In other words, once institutional goals have been established, sets of programs are developed for each goal. These sets of programs are then analyzed to determine which alternative program is most appropriate for the goal in terms of cost and effectiveness.

13. McGivney and Hedges, *Introduction to PPBS*, p. 48.

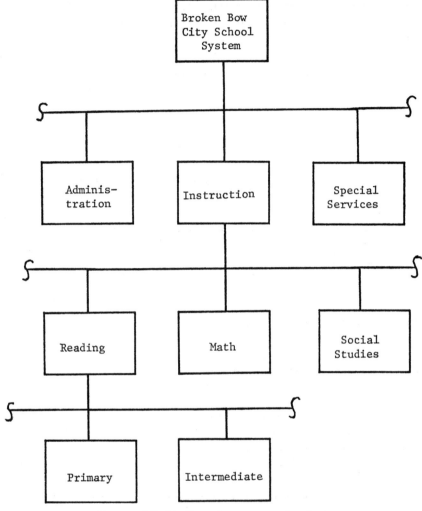

FIGURE 3.2. Portion of a program structure

One process for developing a program is illustrated in subsystem 2.0 of Figure 3.1.

Develop Program Goals (2.1)

The broad goals identified in 1.3 are not specific enough for individual program use. More specific goals, goals which give greater focus, must be developed by the professional staff as a first step to program development.

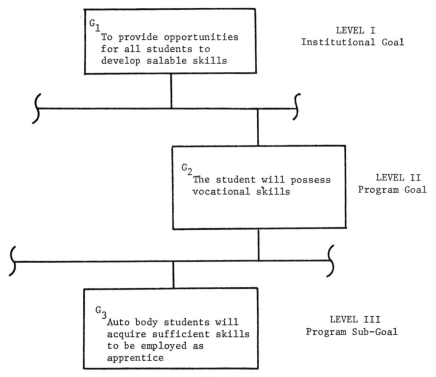

FIGURE 3.3. Hierarchy of goals

These more specific goals should, in some way, relate to the general goals. Each program goal should represent a general goal (1.3); it should specify a student-related outcome; it should be observable; and it should be stated positively. An example of this relationship can be shown as a hierarchy of goals (see Figure 3.3). Notice that there are three levels of generalization in the goal hierarchy. The G_1 level resulted from the goal study (1.3), whereas G_2 and G_3 level goals should be developed by the professional staff. Observe that goal G_3 does relate to G_2, and G_2 relates to G_1. The G_3 goal then serves to give direction to subsequent subgoals and program objective development: the identification of program content and activities, and evaluation.

Develop Program Outcome Objectives (2.2)

A program outcome objective (POO) is "a desired accomplishment that can be measured within a given time and under specific conditions."[14] The

14. *Planning, Programming, Budgeting System Manual*, 2d preliminary ed. (California State Department of Education, 1970), p. 9.

attainment of a program-outcome objective enhances the realization of institutional goals provided the objective "modifies" program goals. Program-outcome objectives must possess the following characteristics:

1. The POO must describe the learner.
2. The POO must describe the learning situation.
3. The POO must describe what is to be learned.
4. The POO must describe the level of performance that is acceptable.

An example of a program outcome objective for G_3 of Figure 3.3 might be:

G_3—Auto-body students will acquire sufficient skills to be employed as apprentices.

POO—Ninety percent of all students completing the requirements of the two-year auto-body repair program will secure part-time employment as apprentice auto-body repairmen within three months of program completion.

Develop Content Area for Program (2.3)

Once institutional direction has been established (goals), program goals developed, and outcomes specified (program-outcome objectives), content can be determined. Utilizing the example given above of the auto-body program, content areas and learning units can be developed for each program-outcome objective. For each content area, it might be necessary to develop area goals and more specific objectives. Some educators refer to the area or unit objectives as behavioral objectives.

Determine Instructional Approaches for Programs (2.4)

One attribute of a professional educator is the ability to draw from a broad repertoire of instructional approaches. The professional educator should be able to employ whatever instructional approaches are most appropriate for the particular goals, the outcomes anticipated, and the content explored. Here one might consider such approaches as individual laboratory experiences, small-group discussion, large-group lecture, independent study, project work, team teaching, or a multiplicity of other kinds of instructional approaches. The identification of the instructional approaches is necessary for determining all administrative arrangements.

Design Measures of Program Effectiveness (2.5)

Measures of program effectiveness are related to two considerations: the projected outcomes of the program and the program costs. Alternate programs developed for single goals, with projections of program effectiveness, will provide important information for the selection of the best programs. The selection will be based upon maximum effectiveness and minimum cost.

Determine Administrative Arrangements (2.6)

Unfortunately, many school systems build buildings, order equipment, hire staff, and secure materials with only limited knowledge of what it is that the school system is going to do. All too often the buildings and equipment are totally inappropriate for the programs that evolve. Personnel find themselves teaching in fields for which they are poorly prepared, and the materials available are in need of modification before they can be used. Certainly, decisions about administrative arrangements must have a relationship to all other considerations in the program-development process.

Each individual program will require the consideration of certain administrative arrangements. By the very nature of some programs, school policies and regulations may need modifications. In some instances, new policies may be needed before the program can be implemented. Policies relative to personnel required for certain types of programs, use of facilities and equipment, class time schedules, etc. may need revision before program installation can be realized.[15]

Along with policy considerations, other kinds of determinations must be made in 2.6. These include: the identification of materials and equipment (2.6.1 and 2.6.2); a description of facilities (2.6.3); and the identification of needed personnel (2.6.5).

Develop Program Budget (3.0)

The actual program budget reflects just the direct cost of the program and subprograms. Cost categories used in the individual program budget are most often identified on budget forms provided by the business office. These forms are then used to project program budget for the budget year.

15. Montello, "Systems Planning for Higher Education," p. 141.

The budget should provide for making projections of future costs. Not only should decisions of projected costs for the budget year be provided for, but a multiyear projection should be made to facilitate future planning efforts. By including a multiyear budget, educational planners can anticipate future funding needs.

Develop Evaluation Design for Programs (4.0)

The evaluation design should reflect a process for determining whether what was planned and stipulated was accomplished. The evaluation process should incorporate procedures for monitoring program implementation as well as assessing the achievement of objectives. The evaluation design must reflect a rational process for determining the level of achievement of the program-outcome objectives and for reporting the results of the evaluation.

Implement Program (5.0)

The process of program implementation requires prior approval by the board of education (5.1). Once this has been acquired, a systems methodology like PDM or CPM (5.2) can be utilized to develop an implementation plan. This plan will take into account the actual running of the program (5.3) and its evaluation (5.4).

Summary

PPB is a totally integrative process of comprehensive curriculum development whereby alternative courses of action are developed for identified goals. Courses of action that provide for maximum effectiveness with minimum utilization of resources are selected for implementation. The PPB process must be specially designed for each individual school system. It facilitates decision-making through a rational process of organizing data and information important to specific decisions.

MANAGEMENT INFORMATION SYSTEMS

PPBS is a type of Management Information System (MIS). By becoming familiar with PPBS, the reader will have a better understanding of Management Information Systems in general. Therefore, as each area of an MIS is given and explained, the reader should refer back to the preceding section and attempt to relate the idea of MIS to PPBS.

An MIS will probably be unique for any given organization. Even in a small organization, the amount of data and information to be processed, used, and understood can be phenomenal. Thus, a system is needed that will bring all relevant information to bear on both day-to-day decisions and special problems. MIS is, in a way, a form of organized data and information flow. According to Martin, a system is "simply an assemblage or combination of things or parts forming a complex whole." [16] Systems management emphasizes the relationships between components, how and why these components interact, and methods of controlling and changing these relationships.

Many confuse the terms *system* and *computer*. They are not the same. In fact, a computer is a system, but a system is not necessarily a computer; that is, a computer is only one of many systems. Other systems that are well known are the solar system, the telephone system, and the local electric-power generation and delivery system. All of these were in operation long before the advent of the modern electronic computer. An MIS does not necessarily require a computer. An example is PPBS, which can be made completely workable by conventional information accumulation and distribution techniques.

The main purpose of an MIS is to inform management when action is needed to give alternative courses of action. This leads to management by exception, or "red-flag" management.[17] Ordinary day-to-day decisions are handled in a routine fashion, but when there is a deviation from the norm, management is informed, given all relevant information available, and given several possible solutions. This enables management to make a quality decision in such instances. Information can be used for planning, controlling, and directing—that is, for decision-making. Planners decide what needs to be done and what will be done. Controllers and directors assess the on-going plan, determine deviations, and decide what must be done to keep the plan in effect.

Data and Information

The terms *data* and *information* have both been used in the discussion so far. What is the difference between them? One definition holds that data

16. E. W. Martin, Jr., "The Systems Concept," in *Systems, Organization, Analysis, Management*, ed. David I. Cleland and William R. King (New York: McGraw-Hill, 1969), p. 49.

17. David I. Cleland and William R. King, *Management: A Systems Approach* (New York: McGraw-Hill, 1972), p. 426.

are a set of characters or signals to which a significance can be assigned, such as a credit card.[18] According to another definition, data are the un-evaluated facts and messages that exist in the environment, or just raw statements of fact.[19]

Before data can really be useful, they must be changed into informa-tion, which is defined as data selected in light of the problem at hand. Data can be changed into information. Also, facts that are data in one situation may be information in another. For example, a credit-card number plus a sale price is information that tells a salesclerk what to do. However, the same facts are data to the billing clerk, who needs to consult the customer ledger for additional data in order to get the proper billing information.

One way of looking at the transfer motion of data into information is called the black-box concept (see Figure 1.1). The input is data, and the output is information. Something has to happen (process) to make the input useful to users in terms of output. For example, the telephone is a processor. Electronic data flows in, is transformed, and comes out as speech information.

By modifying the black box with the addition of a sensor, the system can have feedback. This enables the processor to be self-correcting (see Figure 3.4). A common example of feedback would be the central heating system in a home. The cool air input is heated (processed) and delivered to the space being heated; a thermostat (sensor) measures the output data and sends information back to the processor, instructing it as to how to modify the output.

Information, in order to be useful, must possess certain attributes. Hussain lists the following four:

1. Timeliness. This depends to a great extent upon the user's needs. Some require certain information quickly, while others do not. Obviously, conflicts might

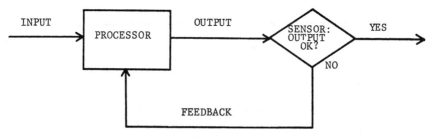

FIGURE 3.4. Concept of sensing and feedback in a system

18. K. M. Hussain, *Development of Information Systems for Education* (Englewood Cliffs, N.J.: Prentice-Hall, 1973), p. 81.
19. Cleland and King, *Management*, p. 415.

arise with multiple users of the same information. It is important that the right information be available at the critical time, so timeliness is an extremely important attribute.

2. Accuracy. This, in simple terms, means the absence of error. There are degrees of accuracy, however, so a literal interpretation is not in order. For example, in business one might round prices to the nearest cent. Accuracy can be measured in terms of reliability, as well. A system which gives the correct information 99 times out of 100, on the average, can be said to be 99% reliable. The user must determine whether or not this is sufficient. Faculty paychecks ought to have 100% reliability in accuracy, but the room thermostat could vary one or two degrees Fahrenheit with little problem.

3. Errors. These can come from improper input, faulty procedures, incorrect processing rules, and equipment malfunctions. For example, the modern electronic computer's peripheral equipment can be improperly used, i.e., improperly punched cards, cards out of order, poor programming, and card-reader mechanical failure. As errors increase, accuracy decreases.

4. Relevancy and completeness. Information must be related and applicable to the problem at hand. Timely, accurate, but irrelevant information is of little use. When all that is relevant is included, it is said to be complete information. Some systems designers generate a considerable amount of irrelevant information in an attempt to be complete. This generally indicates a lack of knowledge about the user's real needs.[20]

An additional attribute of information is *obsolescence.* Information needs tend to change through time; thus, information that is relevant today may be useless tomorrow. The system requires periodic updating in order to generate the proper information. PPBS, for example, is not static. It must be continually inspected to generate and maintain the needed plans, programs, and budgets.

Information can flow multidirectionally; it can flow upward, downward, and laterally. In an ideal situation it flows by the shortest path between the generator and the user. The "chain of command," if it is rigidly enforced, can serve as a block to good information flow, because lateral information flow is restricted. Many organizations operate with little or no upward information flow, that is, from the lower levels of management to the top. Oddly enough, lower-level personnel often have the information required, or at least the data, and are unable to provide it because of organizational blockage. Some methods by which information can flow are word-of-mouth, written correspondence, computer printouts, accounting documents, reports, receipts, purchase orders, time cards, lesson plans, charts, maps, graphs, and drawings.

20. Hussain, *Information Systems for Education*, p. 87.

Determining System Requirements

Cleland and King suggested that the following eight questions be answered in preparation for determining system requirements.[21]

1. What must be known?
2. Where can the data be obtained?
3. Who will gather the data?
4. How will the data be gathered?
5. Who will analyze and interpret the data?
6. How will the analysis be stored?
7. How can the analysis be delivered at the time and place it is required?
8. What kind of protection, or security, is needed?

Obviously, there are two points of view. The users have a need for information, while the systems designer is to provide the information. The user's needs should be the designer's requirements.

An information system should be specially designed for an organization. Goals and objectives, therefore, must be known ab initio. Management is sometimes reluctant to spell out objectives despite the direction that goals imply, but in order for the system to work, these objectives must be known by the designers. Policy must also be clearly stated. Management cannot expect an MIS to be better than the objectives (requirements) specified at the outset.

In the linear programming example that was discussed in Chapter 2, the overall objective was to bus students in order to satisfy the board's desire for better racial balance. There were constraints involved: the number of students of each race required at each school, and the distances to be traveled. If these were not clearly identified for the systems designer, he might interject constraints of his own, perhaps with disastrous results. For example, assume that a management consultant was hired to design a garbage-collection system, including an incinerator, for a large city. Assume further that the consultant put one of his young and gifted assistants on the project. After several weeks he designed the system. It was beautiful—all truck distances were minimized by locating the incinerator in a certain area. Trucks were routed off main thoroughfares in order to speed morning traffic. Other excellent features were also incorporated into the system. However, the proposed incinerator was located within two blocks of the home of the chairman of the aldermanic finance committee—a location that was certainly out of the question. The assistant could not be held at fault since no one had told him that certain locations were out of bounds

21. Cleland and King, *Management*, pp. 425–26.

for political reasons. He did the best he could do with incomplete information. This points up the fact that he needed all the relevant (or complete) information in order to do the job well.

Designing the System

An actual system design cannot begin until the requirements (objectives) of the system, along with an idea of the basic methods of data collection and integration, are known. Cleland and King suggested a six-step design process. [22]

1. Analyze goals and objectives.
2. Develop a decision inventory.
3. Analyze information requirements.
4. Develop a data base.
5. Analyze software requirements.
6. Analyze hardware requirements.

The second step, development of a decision inventory, means that the decisions to be made within the organization must be known, just as goals and objectives must be clearly understood. The decision inventory allows information to be generated that will be useful in the management process. This is, after all, what the MIS is for. Step by step, a decision that must be made would be stated, and then the particular information needed in order to make the decision would be stated. Development of the data base would follow, and the base would allow all stated information requirements to be met. In order to develop a data base a data-collection system must be employed.

The traditional approach to data collection, and also the oldest approach, is to analyze the current system.[23] This is probably the origin of the title "systems analyst." A data-collection system, although it may be cumbersome, does normally exist. The traditional approach is often time-consuming and expensive.

The so-called innovative approach is an alternative method of data collection and subsequent design.[24] The designer generates a new system, paying little or no heed to existing data-collection processes. Although this may cause a need for policy and procedure changes, the approach is generally quicker and sometimes less costly than trying to adopt the existing data-collection system.

22. Ibid., p. 429.
23. Hussain, *Information Systems for Education*, p. 233.
24. Ibid., p. 236.

A combination of these two methods of data collection is often used. The combined approach does not guarantee any better results, but may improve the time frame of the traditional approach, and reduce costs without forcing radical change in current policies. Implementation requires that an innovative outline be determined and compared with the existing system. Problems of obvious conflict are then resolved, preparations are made to complete the design, and the design is finally completed.

In determining system requirements, it is important to realize that the data collection must be integrated. The various components must be compatible and interlinked. This generally results in what is called an "integrated data base."[25] Data are recorded in their simplest form so that they can be used in a wide variety of ways. Data should normally be indecomposable, that is, such that they cannot be broken down into smaller components.

Various techniques can be used to collect data, such as interviews, questionnaires, Delphi techniques, observation, and historical records. The particular combination of techniques needed is dependent upon the situation.

In designing the "software," the user and the designer must be able to state specifications for operations, hardware, and personnel.[26] With these specifications, the designer identifies characteristics and develops standards which should meet the user's objectives. Hardware must be compatible with software specifications.

Output Specifications

What output does the user desire? What output can be generated by the system? Compromise is often required in order to get an answer to these two questions. Good output can be understood by any user or potential user provided that it has a direct relationship to the goals and objectives of the system.

Input Specifications

Changing input methods after a system is operational can be tedious. The user and the designer must consider what may be needed in the future and attempt to put data into the system accordingly. It is easier to aggregate indecomposable data than to break decomposable data into smaller

25. Cleland and King, *Management*, p. 420.
26. Hussain, *Information Systems for Education*, pp. 256–66.

components in the input specifications. A useful rule of thumb is to put data into the system at the lowest useful level and then aggregate as necessary.

Programming Specifications

Should it be deemed necessary to use a computer, instructions for programming are necessary. Capabilities and features needed in the program, along with the input and output requirements and the format desired for output, are all part of these specifications. Sometimes input format is also required.

Documentation Specifications

A written record of the development of the MIS should be maintained for future reference. This is useful for further development, routine evaluation, training, and maintenance. Human memories are fallible; a written record of the development of the MIS and subsequent modifications will assist future users of the system.

Computers in Educational Management

The modern digital computer has had a tremendous impact on the lives of all of us, and it is only natural for educational leaders to want to utilize the great power of the computer in the operation of schools. The computer has already been put to some very obvious uses in the schools. It has been used in scheduling classes, financial accounting, inventory control, grade reporting, and student-record keeping.

The overpowering characteristic of the computer, which makes it so useful, is its ability to remember. As an example, for hundreds of years, accounting data and information were recorded and processed by hand. This involved many man-hours of tedious work. With a computer, all the data can be stored on magnetic tapes, processed into information, and delivered to appropriate users in a fraction of the time previously required. The data stored on hundreds of pages of accounting journals and ledgers can easily be put on one magnetic tape, and the processor can search the tape and extract exactly the desired pieces of data in any predetermined order. These data can be processed into user's information and delivered in printed form.

The computer has several other attributes which should be noted. It is fast, accurate, and virtually untiring. On the other hand, there are certain drawbacks. Computers require detailed instruction in order to operate. In the discussion of MIS, the importance of the user has been stressed. This is also true for the computer; the user must give the systems analyst his requirements (objectives) before the computer can be expected to perform as desired.

Since a computer is a system, it operates much like an MIS. There are input, output, and processing. The input employs some coded device, such as punched cards or magnetic tape. Processing is done by the Central Processing Unit (CPU), which receives the input, processes, allows storage and control of the data, generates information, and translates the output into a usable form. The forms of the output include magnetic tape, printed material, and punch cards.

MANAGEMENT BY OBJECTIVES AND RESULTS

This comprehensive management system, known variously as MBO and MBO/R, and as Results Management, has been used in industry for several years with varying outcomes. Some organizations have been very pleased with the results of MBO/R, and it is interesting to note that the managements of organizations realizing strong success from MBO/R have almost invariably supported and maintained the concept at all levels of the organization. The managements of these organizations have promoted frequent communication sessions regarding organizational objectives, with much attention given to the rationale for such objectives. Moreover, all levels of management and professional personnel play a role in the setting of goals and objectives.

MBO/R has been defined as a set of processes: a generalizable approach that can be adapted to a variety of institutions, including education.[27] George S. Odiorne describes MBO/R as a set of processes:

Whereby the superior and subordinate managers of an organization jointly identify its common goals, define each individual's major areas of responsibility in terms of the results expected of him, and use these measures as guides for operating the unit and assessing the contributions of each of its members.[28]

27. *Management by Objectives and Results* (Arlington Va.: American Association of School Administrators, 1973), p. 2.

28. George S. Odiorne, *Management by Objectives* (New York: Pitman, 1965), pp. 55–56.

MBO/R should be contrasted with management by controls. Those who manage by controls are concerned with what their personnel are doing; those using MBO/R are concerned with results. Educational management is characterized by management of means, or controls. Objectives and results are often unclear, and the focus of management seems to be the means instead of the outcomes of the processes. On the other hand, MBO/R users spell out objectives and the results they expect to obtain, then concentrate on realizing these objectives. One of the pitfalls in MBO/R becomes apparent at this point. Some MBO/R users have become disillusioned because they fail to monitor and control the processes, concentrating exclusively on objectives. Managers must pay attention to the processes per se and at the same time keep the objectives in mind.

Management by objectives and results in education can be thought of as education by objectives and results. This has been defined as "a system of operation that enables the organization and its personnel to identify, move toward, and lock onto objectives as well as to manage more effectively for desired results."[29] Education By Objectives/Results (EBO/R) consists of MBO/R plus Instruction By Objectives (IBO) and management, or

$$EBO/R = MBO/R + IBO/R$$

A good deal of effort is currently being put into IBO/R by various curriculum specialists who are well aware of the fact that education must strive to achieve better results in instructional programs.

A general MBO/R model has been developed by the American Association of School Administrators (AASA), and it is shown in Figure 3.5. Arrows to the outside of the model represent recycling and apply to instances where objectives are not feasible or are inconsistent with goals. Note that once the operational strategy is determined, there is no further recycling until after evaluation and auditing take place. MBO/R users must evaluate and audit in order to determine whether or not they have realized or met their objectives.

In a study of the impact of MBO/R, it was found that:

1. Top-level management should explain, coordinate, and guide the program.
2. Attention must be given to the method of implementation. MBO/R will not implement itself.
3. Each organization must decide on the number and frequency of feedback sessions on an individual basis. Too few sessions are as bad as too many.
4. Problem areas must be recognized—excessive counseling time,

29. *Management by Objectives and Results*, p. 5.

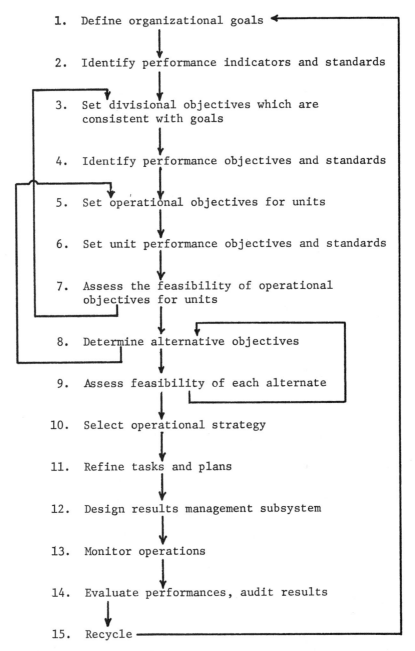

1. Define organizational goals

2. Identify performance indicators and standards

3. Set divisional objectives which are consistent with goals

4. Identify performance objectives and standards

5. Set operational objectives for units

6. Set unit performance objectives and standards

7. Assess the feasibility of operational objectives for units

8. Determine alternative objectives

9. Assess feasibility of each alternate

10. Select operational strategy

11. Refine tasks and plans

12. Design results management subsystem

13. Monitor operations

14. Evaluate performances, audit results

15. Recycle

FIGURE 3.5. General MBO/R model (from *Management by Objectives and Results*, p. 27)

overemphasis on quantitative goals, the tendency of top management to sweep aside negative reaction by subordinates. All these undermine foundational support.

5. Constant review is required to determine effects of the program.[30]

Some critics have called MBO/R "dehumanizing." This need not be the case, if management operates on the basis of McGregor's Theory Y assumptions. These assumptions stipulate that people are to be respected for their contributions, are considered to be essentially self-motivated, and care for the future of the organization. Theory Y assumptions are basic to effective MBO/R.[31]

In conclusion, some comments on objectives are in order. To be usable in MBO/R, objectives must be understandable, performance-oriented, measurable, challenging, realistic, significant, accurate, and brief. Great care must be exercised in the formulation of objectives, for they are the benchmarks against which the educational organization must measure its success.

READING LIST

ALIOTO, ROBERT F., and JUNGHERR, J. A. *Operational PPBS for Education.* New York: Harper & Row, 1971.

American Association of School Administrators. *Management by Objectives and Results.* Arlington, Va., 1973.

California State Department of Education. *Planning, Programming, Budgeting System Manual.* 2nd preliminary ed. 1970.

CLELAND, DAVID I., and KING, WILLIAM R. *Management: A Systems Approach.* New York: McGraw-Hill, 1972.

CURTIS, WILLIAM H. *Educational Resources Management System.* Chicago: Research Corporation of the Association of School Business Officials, 1971.

FARMER, JAMES, *Why Planning, Programming, Budgeting Systems in Higher Education?* Boulder, Colo.: Western Interstate Commission for Higher Education, February 1970.

FISHER, G. H. "The Analytical Bases of Systems Analysis." *Systems, Organization, Analysis, Management.* Ed. DAVID I. CLELAND and WILLIAM R. KING. New York: McGraw-Hill, 1969.

HARTLEY, HARRY J. *Educational Planning-Programming-Budgeting: A Systems Approach.* Englewood Cliffs, N.J.: Prentice-Hall, 1968.

HUSSAIN, KHATEEB M. *Development of Information Systems for Education.* New York: Prentice-Hall, 1973.

30. John M. Ivancevich et al. "A Study of the Impact of Management by Objectives on Perceived Need Satisfaction," in *Fundamentals of Management: Selected Readings* ed. James H. Donnelly et al. (Austin: Business Publications, 1971), p. 145.

31. *Management by Objectives and Results*, p. 9.

IVANCEVICH, JOHN M.; DONNELLY, JAMES H.; and LYON, HERBERT L. "A Study of the Impact of Management by Objectives on Perceived Need Satisfaction." In *Fundamentals of Management: Selected Readings*. Ed. JAMES H. DONNELLY et al. Austin: Business Publications, 1971.

KAUFMAN, ROBERT A. *Educational System Planning*. Englewood Cliffs, N.J.: Prentice-Hall, 1972.

McGIVNEY, JOSEPH H., and HEDGES, ROBERT E. *An Introduction to PPBS*. Columbus: Charles E. Merrill, 1972.

MARTIN, E. W., JR. "The Systems Concept." In *Systems, Organization, Analysis, Management*. Ed. DAVID I. CLELAND and WILLIAM R. KING. New York: McGraw-Hill, 1969.

MONTELLO, PAUL A. "Systems Planning for Higher Education." In *Accountability: Systems Planning in Education*. Ed. CRETA D. SABINE. Homewood, Ill.: ETC Publications, 1973.

NOVICK, DAVID. *Origin and History of Program Budgeting*. Santa Monica: Rand Corporation, 1966.

ODIORNE, GEORGE S. *Management by Objectives*. New York: Pitman, 1965.

CHAPTER 4

A System for Comprehensive Program Planning and Implementation

There are many texts on systems and systems analysis in education. Unfortunately, very few of these relate theories and techniques directly to educational applications. It is, therefore, the purpose of this chapter to utilize the techniques already presented by applying them to an actual school-system setting.

The following discussion will relate the application of some of the systems techniques discussed in Chapter 2 to an actual educational setting. A LOGOS model will be used to illustrate a program initiated by the faculty of the Department of Educational Administration, Georgia State University, for an in-service degree program in Clayton County, Georgia. This program was developed for, and implemented with, a group of twenty-five school-level administrators in the Clayton County Public School District. The program culminated with the awarding of the sixth-year specialist's degree in educational administration.

BACKGROUND FOR THE MODEL PROGRAM

It was the purpose of the departmental faculty to explore a system of new instructional modalities for upgrading the managerial and leadership competencies of school-level administrators. Traditional programs of administrator preparation have been subject to extensive criticism over the last few years for their lack of relevance to the real world of public school administration. It was the intention of the faculty to develop a program which would have direct application to the leadership problems and managerial dilemmas that administrators face on the job from day to day.

The faculty envisaged its involvement in the training of educational leaders as bringing about a fusion between new knowledge relative to the management and leadership processes and the problems routinely encountered by public school administrators.

Over the years the design of university programs for preparing educational administrators has not been directly related to the exigencies confronting administrators. Most often pre-service and in-service programs have been academically oriented, and experiences were selected from a cafeteria of university courses. Mainly, learning was primarily vicarious in nature, with few real-world experiences for students.

Students in these programs have been exposed to the many theories, principles, and procedures for operating schools. Designers of such programs assumed that this generalized information could and would be applied to actual situations on the job.

It was of utmost importance to this new program that knowledge not be eliminated from training experiences, but be made operational in actual administrative settings. With this consideration advanced as a premise, the role of the faculty of the Department of Educational Administration was redefined as that of finding real-world educational settings where actual problems might be utilized as the medium of learning.

Using this newly defined role, the faculty viewed the administration of schools along two dimensions. These two dimensions have been outlined by Kenneth E. McIntyre of the University of Texas as *administering* and *improving*. He used these two terms deliberately to emphasize two aspects of the administrator's responsibility, *maintenance* and *change*. Both maintenance and change are necessary functions in all instructional programs. Unfortunately a principal's time is usually spent dealing with the former. All too often the administrator has little time to do more than plan for staffing, arrange for instructional space, order materials, complete routine reports, and the like. These kinds of activities consume the day of the administrator to the extent that very little time or energy is left for improving instruction. Rarely do administrators involve themselves in the kinds of activities that enhance or improve instruction.

Planning for the project for Clayton County administrators involved the faculty of the department in an exploration of concepts, principles, and practices that could be employed in job-related learning experiences. The intent of the faculty was to provide job-related experiences that would assist the participating administrators in making better management decisions for many kinds of activities that occupy most of their time. Beyond management-decision skills, the faculty endeavored to assist the participants in the development of competencies for implementing the many processes necessary for improving the instructional programs of the schools and the system as a whole. In this regard, a field-based competency program was developed and implemented in the school system.

MODEL FOR A FIELD-BASED IN-SERVICE PROGRAM

The reader should recall that the LOGOS technique is a modeling methodology for graphically depicting the total relationships of all elements in a system. Figure 4.1 illustrates the LOGOS model for the field-based in-service program for school-level administrators of Clayton County, Georgia.

Basic to the model are two major subsystems: "Develop Skills in Management Competency Areas" and "Develop Skills in Leadership Competency Areas," 1.0 and 2.0, respectively. The first relates to all of those competencies which have to do with the maintenance of the status-quo; that is, the ordering of books and materials, making schedules, collecting and disbursing monies, facility maintenance, and so forth. On the other hand, leadership competencies relate to those activities which are necessary to promote and support desired change. This may involve a principal in goal-setting activities, building communications networks, planning in-service training, and monitoring and evaluating programs.

Develop Skills in Management Competency Areas (1.0)

The initial phases of the program involved the principals in numerous group activities which focused upon the identification of management competency areas (1.1). The process for identifying these areas included group-building exercises for the establishment of an esprit de corps. These activities were planned to enable individuals in the group to realize a high degree of mutual trust and acceptance. Once this was achieved, the energies and expertise of individual group members could be readily used in the identification of needed management competencies.

The concepts of decision-making and time-management were identified by the group of administrators as areas of paramount importance (1.1). This was accomplished through a group process involving all twenty-five administrators. Competency-development experiences were then planned for these two areas as well as for other important competency areas (1.2).

Decision-making

A number of individual and small-group decision-making exercises were engaged in by the participants. These exercises included in-basket simulation and decision games.

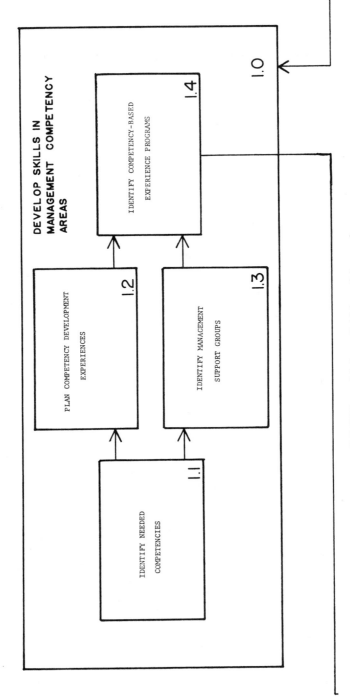

FIGURE 4.1. LOGOS model for a field-based in-service program

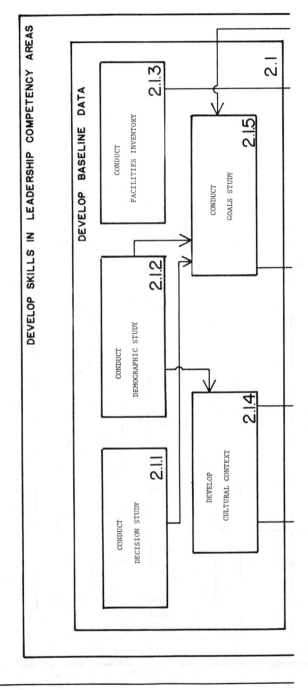

FIGURE 4.1. LOGOS model for a field-based in-service program (continued)

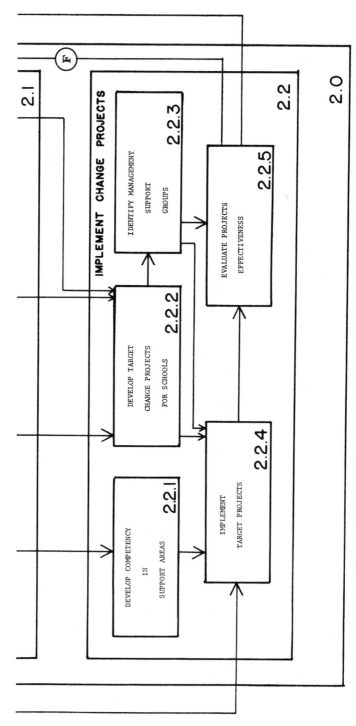

FIGURE 4.1. LOGOS model for a field-based in-service program (continued)

90

In some instances, when group decision-making was the focus, a video-audio tape was made of the small groups in the actual decision-making exercise. Analyses of individual and group effectiveness were conducted through the utilization of several observational instruments. An individual group-role instrument was employed in an effort to provide a system for identifying the types of individual roles played by each participant during the discussion. Additionally, the Bales twelve-category interaction-analysis system for analyzing group problems was used for determining the phases and patterns of group problem areas. The outcome of these analyses provided information that the participants used for gaining insight into and understanding of the processes of decision-making at both the individual and group levels. In order to facilitate subsequent decision-making activities, participants were encouraged to practice various role behaviors and group interventions that were found to be effective.

Time Management

Time management was another serious problem confronting the administrators. Efforts to assist the participants in the development of skills for the effective utilization of time were initiated through the maintenance of daily time logs for a period of one week. The information from these individual time logs was then classified by the participants into general areas of work activity. This was followed by a time management workshop where a system for analyzing time logs was presented to the group. This was followed by a presentation on the development of strategies for better time utilization. Basic to the analysis and strategy development were individual answers to the following questions: Is the task necessary? Can it be done as well or better by someone else? Can it be routinized? Does it reveal symptoms of a greater problem?

Other areas of school management were included as primary interests and explored in the first phase of the program. Legal implications for principals, policies and regulations, and school building maintenance were investigated. Several instructional techniques were used.

An in-basket simulation exercise was employed for the exploration of legal implications for principals. Once the particpants had completed the in-basket simulation and had spent several hours discussing possible alternative courses of action in the disposition of each in-basket item, a list of legal questions was developed relative to the in-basket items. The participants directed these questions to an expert in the field of school law. The legal expert addressed himself to the many legal ramifications of the questions that were raised in the in-basket by citing arguments of current cases being adjudicated in the courts. Additionally, recent court decisions related to corporal punishment, discipline, negligence, nonrenewal of teachers, and due process procedures were discussed. The workshop was concluded with a discussion of legal guidelines for principals.

Other Areas of Concern

The principals had other important concerns relative to managerial competencies. These included:

1. Pay and fringe benefits
2. Training programs for custodial staff
3. Supervision of custodians
4. Teacher–custodial staff relations
5. Coordination of building maintenance
6. Upkeep of campus grounds
7. Planning for custodial services

Numerous workshops were planned and implemented for these areas, using university staff-members who had expertise in the areas. The development of skill in management-competency areas identified by single program participants was gained through the utilization of the Management Support Group.

Management Support Group (MSG)

It became apparent to the instructional team from working with the group of administrators that there were many individual problem areas that needed immediate attention. In this regard, management support groups were organized to give direct assistance to individual administrators. MSGs of two and three university personnel with specific areas of expertise were formed for the sole purpose of working directly with individual administrators in the resolution of identified problems. The MSGs conferred directly with the individual administrators and assisted them in the delineation of problems and the implementation of the alternative that was selected as being the most appropriate. The concept of the MSG will continue throughout the project since a cadre of expert university personnel is readily available to the school system's administrators so that school problem areas can receive immediate and expert attention.

Generally, the MSGs were organized as disposable organizations. When solutions to problem areas were found and implemented, the MSG was dissolved. Members of MSGs were identified in such a way that specific areas of expertise could be employed in the resolution of problems. It was not uncommon to find university personnel with expertise in guidance and counseling working closely with experts in professional staffing and curricular innovations.

Develop Skills in Leadership Competency Areas (2.0)

Instructional leadership is certainly an important activity of the principalship. In this context, *instructional leadership* and *change* are synonymous. Change implies altering the status quo or moving the system out of equilibrium. Under this condition, the impact of such change on the many constituencies of the system must be taken into account. Subsystem 2.0 of Figure 4.1 is a model for making instructional modifications. It could be called a "change model."

Basic to the model is the assumption that the school-community should be involved in the goal-setting process. Moreover, new program directions are more likely to be accepted if the school constituencies are involved in the setting of goals.

The Clayton County program provided an opportunity for the principals enrolled in the program to gain skill in the leadership competency area through the implementation of subsystem 2.0 of the training model. The implementation of subsystem 2.0 necessitated the establishment of numerous task forces made up of the participants in the training program. Their involvement in this implementation process provided an opportunity for them to gain in leadership competency areas and to install a change model in their individual schools.

One of the first elements of the system implemented was "Conduct Demographic Study" (2.1.2). To initiate this study, the participating administrators were divided into five task forces. One task force was identified for each of the four high school attendance areas, including the feeder elementary and junior high schools, and a fifth for the total county. Each task force conducted a demographic study for its constituency.

The purpose of the study was to identify important population, employment, civic and religious, geographic and housing, income, and business and industry data about the communities. These data were then used as a basis for subsequent instructional development.

Included below is an excerpt from the report of one of the demographic studies.

Clayton County is located in the north central section of the state of Georgia. It borders the city limits of Atlanta to the south.

Prior to World War II, Clayton County was a small sparsely populated, agricultural community. The major transportation artery was the Southern Railway line which passed through the county. After World War II, Clayton County began to flourish as a suburban county becoming part of metro Atlanta with its industrial developments, airport expansion, and bedroom space for Atlanta's working population. The rapid increase in population has produced a tremendous boom. This has been characterized by inflating land values, road construction, subdivision developments, massive water and sewage development, and the addition of many other services the county citizens have had to pay for with higher taxes.

Jonesboro is the county seat of government and still has a large amount of undeveloped land south of its city limits. It is located in the southern part of Clayton County and has the second largest school enrollment in the county.

The high school attendance area of Jonesboro is served by 7 elementary schools and 2 junior high schools. There is another junior high school under construction and scheduled to open in the Fall of 1974.

In examining the population trends for the Jonesboro attendance area it is evident that the area has grown very rapidly during the past twenty years. It has experienced the greatest per cent of increase in school age population during the last 10 years of all areas of the county. The Jonesboro attendance area grew from an enrollment of 2,004 in 1950, to 7,578 in 1970, and the projected enrollment for 1980 is 14,215.

The population is expected to be over 50,000 in the Jonesboro area by 1980, which will be second only to Morrow's 60,000. The in-migration has doubled in the past 10 years along with the birth rate. Moreover, the in-migration is expected to more than double in the next ten years.

The characteristics of the population of this area are much the same as the other high school districts of the county. The educational attainment varied very little from area to area, and the age group distribution has remained relatively constant throughout the county. Educational attainment for the Jonesboro area was 8.8 in 1950, 10.5 in 1960, 11.8 in 1970, and is projected to be 13.3 in 1980.

The 25–64 age group has shown the greatest increase in population from 1950–1970 in this area as well as the other areas of the school system. This was attributed to the large in-migration for the county over this same period of time.

The change from rural to suburban has of course brought about drastic changes in the residential patterns as well as the economic patterns in the Jonesboro area. The population density is less in the Jonesboro area than any other in the county. This is due to the large amount of undeveloped land in the panhandle section of the county.

Most of the housing is single family, but there has been a great increase in multi-family housing during the past 2 to 3 years. It is projected by 1980 that this area will have the greatest percentage increase in multi-family housing.

In Clayton County there is a large percentage of residents who are employed in other counties. The Jonesboro area has approximately 45% of its working population employed in Clayton County. Compared to the other areas in the county, this is the largest percentage of residents in a particular part that are employed within the county. The approximate overall county percentages are 37% working in the county and 63% working outside the county.[1]

Concurrent with the demographic study, a decision study was conducted (2.1.1). The implementation of this study found the participants organized into a different set of task forces. Each task force conducted a series of interviews with leading citizens of the community. The interviews focused upon the gaining of insight into the major issues confronting the community and the leadership surrounding these issues. The results of this study revealed a rank-ordered list of the twenty-five most influential citizens of the community. In other words, it identified the community decision-makers. These data were then used in the goals study (2.1.5) that was subsequently conducted in the county.

As in the demographic study, the group of participants was organized into five task forces to conduct the goals study. One task force conducted a goals study in each of the four high school attendance areas while the fifth used a countywide constituency.

The Delphi technique was employed in the conduct of the studies. Here it was necessary to identify the many constituencies of the school-community. In addition to students, teachers, administrators, supervisors, and board members, the samples for the studies drew persons from the many constituencies identified in the demographic study as well as the twenty-five community leaders identified in the decision-making study.

Once a sample was determined for each of the five studies, a questionnaire was sent to the members of each sample. The single question the respondents were asked to answer was, "In the next three years, the Clayton County School System should concentrate its energies and resources on: increasing, solving, developing, and preparing _____?" The results from this open-ended question were analyzed and classified into a set of generic statements; for example, all responses relative to the reading program were combined into a single, general statement. This was done in each of the five goal studies.

The second questionnaire listed all the generic statements and requested the same sample to rate each statement relative to its importance on a five-point scale. The analysis of these data provided information for rank-ordering the statements and the elimination of some of the low-priority items.

After the low-priority items were eliminated, a new questionnaire was developed from the remaining items. In this questionnaire the respondents

1. Thomas Rayburn et al., *Educational Targets for the Mid-Seventies: A Study of the Jonesboro Senior High School Attendance Area* (Clayton County (Ga.) Public Schools, February 1974), pp. 3–6.

were provided information about their previous responses to each item and the numeric response that was most frequently used by others in the sample. The participants were then asked to reconsider only those items where they differed in their response from the response that was most frequently offered (consensus). The results of the third questionnaire were analyzed and the items were rank-ordered. Again, low-priority items were eliminated.

Finally, the remaining items were presented to the members of the sample in individual interviews. Here the Q-sort methodology was employed for a final ranking of the items. The final rank-order of items represented the targets that subsequently became the thrust of individual school program-change projects.

Two other elements of the subsystem "Develop Baseline Data"(2.1) of Figure 4.1 involved the group of administrators in research activities. These included the development of an interpretation of the cultural context (2.1.4) and a school facilities inventory (2.1.3). The completion of these two activities finalized the development of the baseline data document (2.1).

The development of leadership competencies for implementing change projects (2.2) is a subsubsystem of the total model. It contains five elements (see Figure 4.1).

At this point in the training program, the participants were instructed to utilize the inputs from 2.1 and make an interpretation of the targets identified in 2.1.5 for their individual schools. As an example, one area study revealed the high-priority target, "Develop program for students with learning disabilities." The task of each administrator was to determine whether that target had meaning for his school. Specifically, he had to implement procedures for determining the need for such a program in his school.

Once the target was identified as primary for a particular school, the faculty of the Department of Educational Administration worked with the principal and his staff in the development of a project appropriate for the target. The general outline that was used for each project included:

1. Statement of the target
2. Appropriate goal statements for the target
3. Goal indicators
4. Project objectives (must modify goal statements)
5. Sets of alternative project activities
6. Identification of resource needs for each set of alternative project activity
7. Outline of methods for evaluating project

In addition to serving as a consultant to the principals on their projects, the departmental faculty performed other facilitative functions.

Specifically, each of the projects required expert inputs not available in the department. It therefore became necessary for the faculty to identify individuals with expertise to assist in the development and implementation of the projects. These experts, together with members of the department and central-office staff from Clayton County, worked as management support groups. Their function was to lend assistance to the principal in the development, implementation, monitoring, and evaluation of the project. This temporary organization provided the principal with ready access to a group of individuals with a wide range of expertise for each project.

Throughout the development, implementation, and evaluation of the projects, the participating administrators were involved in other learning experiences appropriate for developing competence in support areas. The psychological, sociological, and philosophical foundations appropriate for the target projects were explored. This involved the use of faculty members from other departments of the School of Education.

The feedback loops from 2.2.5 provided a mechanism for assessing the overall effectiveness of each administrator. Here the faculty looked at individual project problem areas. This provided information on administrators' competency deficiencies. The faculty then developed experiences appropriate to remediate these deficiencies.

In summary, the first part of this chapter outlined an actual LOGOS model. The conceptual model was explained in detail so that the reader could gain a deeper understanding of the modeling technique and its uses. The following section of this chapter will illustrate the use of PDM for implementing and monitoring the program.

PDM NETWORK FOR IMPLEMENTING A FIELD-BASED IN-SERVICE PROGRAM

The PDM network for the field-based in-service program in Clayton County is illustrated in Figure 4.2. This figure presents the complete network in a generalized form so that the reader will be able to see the total relationship of the major program activities.

Figure 4.3 is a detailed representation of Figure 4.2. It includes all early start, late start, early finish, and late finish times as well as floats, activity duration times, and the critical path. Although Figure 4.3 represents the total PDM network for the program, some minor activities were omitted to preserve clarity of the illustration.

Time durations were established on the basis of previous experiences with similar activities. Moreover, they were considered deadlines for monitoring the purposes of the program.

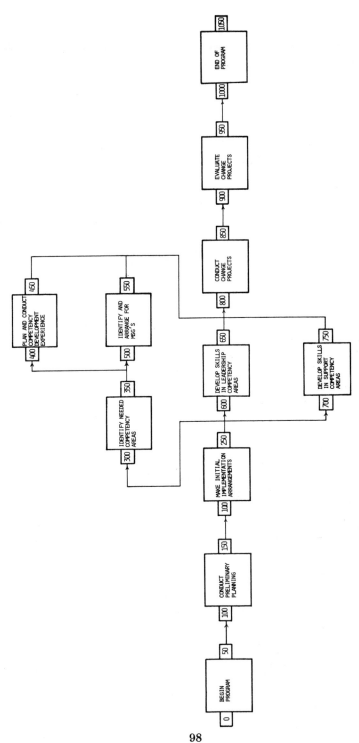

FIGURE 4.2. PDM network for implementing field-based in-service program

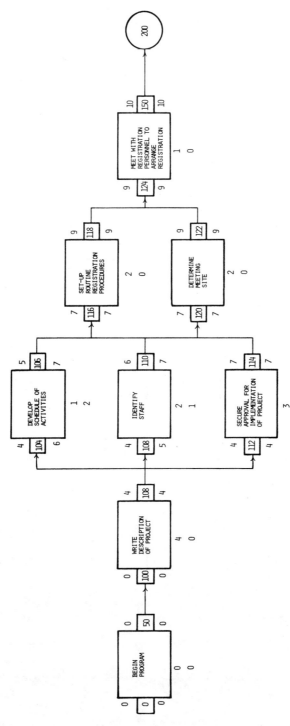

FIGURE 4.3. Comprehensive PDM network for implementing field-based in-service program

99

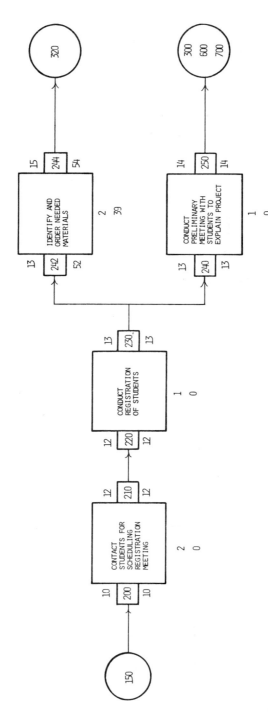

FIGURE 4.3. Comprehensive PDM network for implementing field-based in-service program (continued)

100

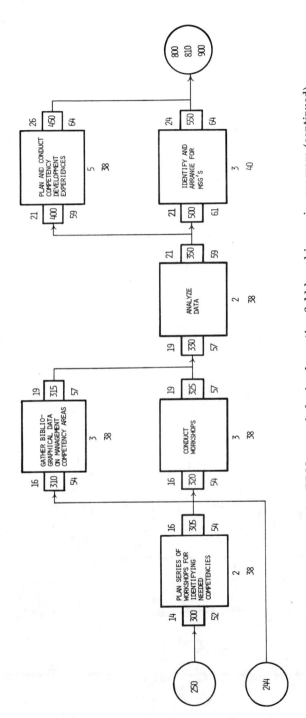

FIGURE 4.3. Comprehensive PDM network for implementing field-based in-service program (continued)

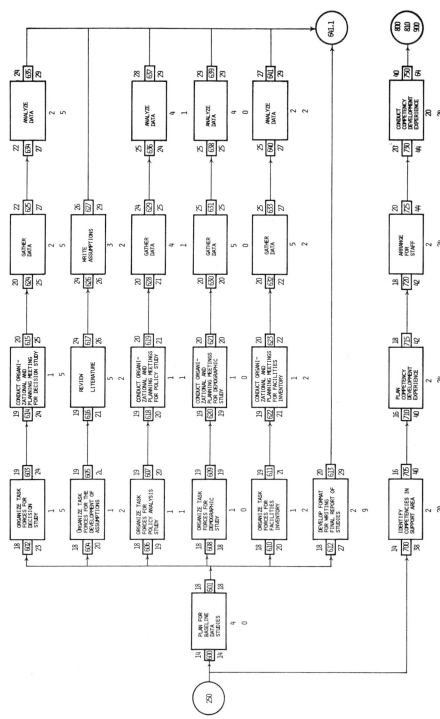

FIGURE 4.3. Comprehensive PDM network for implementing field-based in-service program (continued)

102

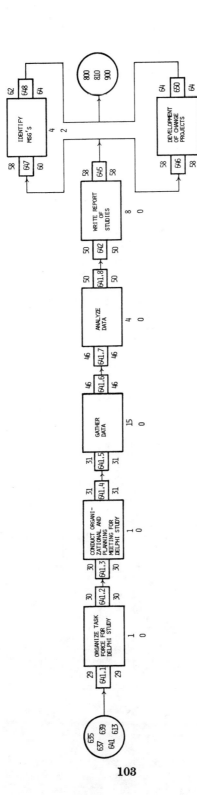

FIGURE 4.3. Comprehensive PDM network for implementing field-based in-service program (continued)

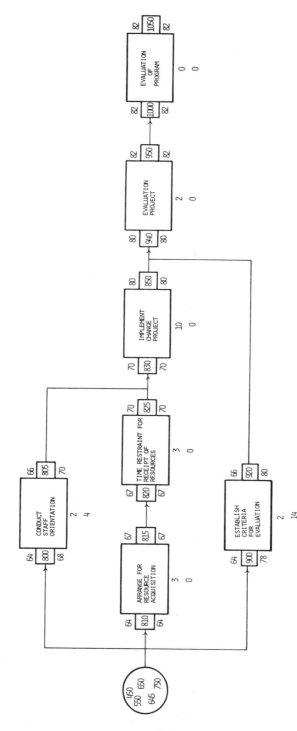

FIGURE 4.3. Comprehensive PDM network for implementing field-based in-service program (continued)

104

Because of the size of the network, it was necessary to continue the illustration on several pages. To facilitate the reader's understanding of the network, circles are used to show that one activity is linked to another on a subsequent page of the figure. This notation technique can be seen by observing that the output of activity 124–150 connects to mode 200 of activity 200–210 on the following page.

The unit of time for the network is a week. From observation one can see that the entire program was completed in eighty-one weeks, or at the beginning of the eighty-second week. The critical path is the path traced by activities: 0–50, 50–100, 112–114, 116–118, 120–122, 124–150, 200–210, 220–230, 240–250, 600–601, 608–609, 620–627, 630–631, 638–639, 641.1–641.2, 641.3–641.4, 641.5–641.6, 641.7–641.8, 642–645, 646–650, 800–815, 820–825, 830–850, 940–950, and 1000–1050.

The use of PDM in planning, monitoring, and controlling the program provided several important kinds of information. It provided a total picture of the interrelationships of all program activities. It provided a reference for all feedback information of the program. Additionally, it facilitated the allocating and scheduling of the many personnel resources of the program.

CHAPTER 5

The Future of Management Systems in Education

Early one mid-March afternoon, the superintendent of a large city school system was interrupted from his review of computerized data relating to staff evaluations by a telephone call from the chairman of the school board. The board chairman had just received important information from a member of the House Education Committee, who had informed him that the state minimum foundations program for education was under review for modification and that a proposed piece of legislation would be forthcoming. The board chairman stipulated to the superintendent that changes in funding were being considered and proceeded to give him the details of the contemplated changes. The board chairman was very concerned about the impact of these proposed changes on the financial welfare of the school system. He ended the conversation by asking the superintendent to investigate the matter.

Immediately, the superintendent telephoned his systems analyst and detailed the proposed changes and the board member's request. Within a matter of minutes the analyst delivered the requested information to the superintendent. This was possible because the school system had previously programmed its computer to simulate numerous funding models, and the school district maintained a management information system appropriate for storing data relative to the models. The nearly immediate access to this information provided the superintendent and board members with important information for apprising the legislative committee of the impact of the proposed changes on their school system.

The preceding example illustrates a capability that currently exists; however, very few school administrators are aware of such potential for improving the decision-making process. This lack of awareness does not result from neglect. It is more a matter of not knowing the potentialities of management systems. Moreover, the holders of the purse strings for the educational complex have not made monies available for exploration and development of new educational management systems. Consequently, educators find themselves continuously trying to utilize the results of investments by industry or the military in systems research and development, adapting these noneducational management systems to educational organizations. What prospects, then, are foreseen for the future of management systems in education?

The authors do not have a crystal ball for prognosticating the future. They can only make projections and predictions based upon the current uses of management systems in education, extrapolations of societal changes, and technological developments in government and industry.

Certainly, no one will be startled by the pronouncement that the use of the systems approach in education will increase. One has only to look to a few places where techniques of systems analysis are currently being employed at an increasing rate of acceptability to find this to be true. An important example is the expanding use of PPB systems. Hartley has stated: "It appears that 18 to 20 states are in the process of mandating reporting procedures based on PPBS. I estimate that more than 1,500 local schools are actively engaged in PPBS. Operational usage of PPBS continues to grow." [1]

The current leadership for this movement appears to be centered in certain states, among them California, Florida, Connecticut, New York, Indiana, Illinois, Colorado, Hawaii, and West Virginia. Although PPBS in education is in an embryonic stage of development, the last several years have been marked by remarkable achievements. A number of school systems, particularly in the Northeast, have developed a high degree of sophistication in their implementation processes. Many have developed their own planning and programming models in which alternative programs are being considered on a cost-effectiveness basis.

Management systems like PPBS are becoming more of a necessity for school districts because of the ever increasing size and complexity of educational organizations. As the size of the organization increases, the need for more effective and efficient systems for processing information will emerge. Routine management decisions will be assigned to the computer. In fact, all of the business functions of a school system will become susceptible to systems-analysis techniques.

Specifically, a large majority of school systems currently use computerized financial accounting systems. Student accounting and grade reporting is another area in which computerized systems are becoming

---

1. Harry J. Hartley, "PPBS in Local Schools: A Status Report," *The Bulletin of the National Association of Secondary School Principals* 56, no. 366 (October 1972): 1.

popular. Many school systems operate their total standardized testing programs on a computerized system. The computer is extremely helpful in scheduling students into classes in schools that employ a modular class arrangement. In many places a new schedule is written for each student each week.

Long-range educational planning is gaining acceptance as a necessary function for school systems. Many large school systems staff an office of planning where sophisticated techniques of systems analysis are employed in the conduct of day-to-day activity.

Techniques like those mentioned in Chapter 2, as well as linear modeling and other quantitative methodologies, are utilized for determining enrollment projections and developing desegregation plans, developing inventory and purchasing models, developing systems of competency-based teacher merit plans, determining future budget projections, as well as many other functions. Certainly, these activities give credibility to the belief that education will see an expanded use of systems techniques in the immediate future and, perhaps, well into the extended future.

One of the most important and obviously difficult functions of schools is the maintenance of good communications with their many publics. This function is expected to become more difficult because the number of publics (constituencies) of the schools is increasing. Also, the extension of the current interest in education is projected to increase in importance, particularly as the availability of leisure time increases. These kinds of societal changes will create a need for comprehensive communications systems, that is, systems that can disseminate appropriate information about the schools to constituencies in need of such information and systems for receiving information (feedback) from the constituencies.

Societal changes are taking place at such a rapid pace that the institutions of our system, particularly education, are finding it difficult to adjust. The targets of need always seem to be moving. As a consequence, educational planners are becoming very interested in *futurism*. This concept has emerged out of similar need in industry and government, and many of the techniques and methodologies that have been developed by these institutions are being adapted to educational planning. This is an exciting field that interested readers can explore through an increasing body of literature not only in industry and government but in education as well.

In the final analysis, it appears that societal activity will become more complex. Growing institutions will find it increasingly difficult to maintain a high degree of relevance to a rapidly changing society, and the ability of large organizations to adjust rapidly will be the test of survival. A case in point is the scramble by the "big three" auto-makers to adjust their output of compact and subcompact cars to the scenario of the future created by the oil embargo of the Arab countries. Through the use of management systems, this adjustment will be made in a relatively short period of time.

In a similar sense, education will have to make rapid changes. As evidenced by recent court orders and legislation, the magnitude of such changes will be great. Certainly, improved management systems will be needed to cope with such changes.

Finally, the authors of this book would like to challenge the reader to predict the impact of the energy crisis on the educational complex. How will the American way of life be changed? What impact will this change have on education? Are there systems techniques for dealing with future educational needs resulting from a changing set of conventional societal values?

Glossary

Words set in SMALL CAPS are defined elsewhere in the Glossary

Activity

An exercise of energy or force whereby resources of personnel, material, and time are expended.

Administrative arrangements

A set of arrangements that includes organizational structure, policies, procedures, facilities, equipment, and personnel.

Arrow

Generally, used as a graphic representation of INFORMATION flow. In CPM and PERT, it represents an ACTIVITY.

Black box

A PROCESSOR in a SYSTEM; the user knows INPUT requirements and OUTPUT requirements, but may not be knowledgeable of the process that changes input to output.

Closed system

A SYSTEM that is shut off from most external influences. Closed systems exist primarily in laboratory work.

Computer

An electronic device capable of very fast DATA manipulation and of vast amounts of data storage.

CPM

See CRITICAL PATH METHOD.

Critical path

The longest path in a network arrangement such as PDM, CPM, or PERT. The critical path cannot be shortened, therefore it sets the total time factor for a network.

Critical Path Method

A system for planning, scheduling, and controlling projects. Also called *CPM*.

Cultural context

A set of philosophical beliefs (values) about society, people, knowledge, and processes.

Cybernetics

The correlation of communications and control; how SYSTEM components function together.

Data

A set of characters or signals to which a significance can be attached.

Decision game

A group decision-making exercise where SIMULATION and role playing are involved in the activities of the game.

Decision-making

Deciding what is going to be done in order to attain GOALS; a part of planning.

Delphi technique

A methodology for organizing and sharing informs such as priorities held by members of various constituencies. The process provides for forcing the constituencies to a high degree of consensus.

Demographic

A concept associated with vital and social statistics.

Descriptor

A label that defines an activity in PDM.

Dummy arrow

Used in CPM to indicate precedence, especially in situations where additional clarity is needed.

Duration time

The time required to perform an activity in PDM or CPM.

Dynamic

Active and ever-changing.

Early finish time

The earliest time a PDM or CPM ACTIVITY can be completed.

Early start time

The earliest time a PDM or CPM ACTIVITY can be started.

Educational Resources Management System

A synonym for PLANNING-PROGRAMMING-BUDGETING SYSTEMS. Also called *ERMS*.

Entrophy

In SYSTEMS theory, the unavailable energy at a given point in time.

Environment

The aggregate of surrounding influences.

Equilibrium

A state of rest; a static state.

ERMS

See EDUCATIONAL RESOURCES MANAGEMENT SYSTEM.

Feedback

A term pertaining to new energy or INFORMATION that is recycled to earlier elements of a SYSTEM.

Feed forward

A term pertaining to energy or INFORMATION that is advanced to subsequent elements of a SYSTEM.

Float

The difference between the time available for an ACTIVITY and the DURATION TIME of the activity.

Flow chart

A planning device that incorporates FEEDBACK and uses simple symbols to indicate various processes in the plan.

Functions

Actions being performed or to be performed in SYSTEMS management. A function is often a SUBSYSTEM.

Generalized Evaluation and Review Technique

A combination of flow charting and CPM, useful in cases where FEED-BACK and probabilistic events occur. Also called *GERT*.

GERT

See GENERALIZED EVALUATION AND REVIEW TECHNIQUE.

Goal

A statement of broad direction, purpose, or intent without reference to time.

Hardware

A functional component, such as a COMPUTER. A piece of equipment.

Homeostasis

In systems theory, the time when the SYSTEM is in balance.

In-basket simulation

A MODEL used to predict performance of individuals in a given systems setting. For example, a set of activities similar to those that an administrator might normally perform in some given time period is one form of in-basket simulation.

Information

DATA combined in such a way as to be useful for management purposes.

Input

DATA or INFORMATION being fed into a SYSTEM, which will be processed by the system.

Instructional approaches

The pedagogical processes of education, e.g., team teaching, independent study, lecture, etc.

Integrated system

A SYSTEM wherein the data is INPUT in its basic form, integrated into INFORMATION as required, and delivered on demand. Also called *integrated data base*.

Language

A descriptive set of symbols used in defining and analyzing a SYSTEM. *See* LOGOS, for example.

Language for Optimizing Graphically Ordered Systems

A language used in model-building. Also called *LOGOS*.

Late finish time

The latest time a PDM or CPM ACTIVITY should be completed.

Late start time

The latest time a PDM or CPM ACTIVITY can be started.

Linear programming

A systems technique for determining optimal allocation of scarce resources.

LOGOS

See LANGUAGE FOR OPTIMIZING GRAPHICALLY ORDERED SYSTEMS.

Management information system

A comprehensive management method by which the needed INFORMATION for management is made available in the correct sequence, at the proper time, and in the most useful form.

Management by Objectives

A SYSTEM whereby all management levels identify common goals, determine individual responsibilities, and then use these measures as operational guides. Also known as *Management by Objectives and Results* (MBO/R).

Management Support Group

A disposable organization or group that works on temporary problems. Also called *MSG*.

MBO

See MANAGEMENT BY OBJECTIVES.

Metasystem

A very large SYSTEM composed of SUPRASYSTEMS, SUPERSYSTEMS, SYSTEMS, SUBSYSTEMS, and subsubsystems.

Model

A representation of a SYSTEM that is used to predict the effect of changes in certain aspects of the system on the performance of the system. Modeling is to design a model. Models may also be scaled-down representations of a portion of an existing system.

MSG

See MANAGEMENT SUPPORT GROUP.

Need

The difference between programs desired and existing programs.

Network

A systems term meaning the graphic representation of a model or plan.

Network analysis

Analysis of a graphic model or plan.

Node

The beginning and end points of an ACTIVITY in PDM, CPM, and PERT.

Objective

A desired accomplishment that can be observed and measured within a given time frame.

Open system

A SYSTEM subject to all external influences.

Output

The result of having processed or transformed INPUT.

PDM

See PRECEDENCE DIAGRAMMING METHOD.

PERT

See PROGRAM EVALUATION REVIEW TECHNIQUE

Planning-Programming-Budgeting System

A totally integrative process of comprehensive curriculum development whereby alternative courses of action are developed for identified goals. Also called *PPBS*.

POO (Program Outcome Objective)

A desired outcome that can be observed and measured within a given period of time and under specified conditions.

PPBS

See PLANNING-PROGRAMMING-BUDGETING SYSTEM.

Precedence Diagramming Method

A SYSTEM for planning, scheduling, and controlling projects. Also called *PDM*.

Processor

A systems term meaning the technique, area, or method in which a transformation takes place.

Program

A plan of activities and support services representing a design for attaining program objectives.

Program budget

A financial statement that relates costs to goals, objectives, and programs.

Program Evaluation Review Technique

A NETWORK technique similar to CPM and PDM, but using three time estimates per ACTIVITY. Also called *PERT*.

Program structure

A hierarchical arrangement of programs that represents the interrelationship of activities to goals and objectives.

Q-sort

A technique for sorting concepts or statements (goals) into an approximated normal distribution relative to labels of subjective preferences. For example, a deck of cards, each containing a unique goal statement would be sorted into five piles of cards relative to labels such as strongly agree, agree, neutral, disagree, and strongly disagree.

Quantitative analysis

The use of mathematics and systems techniques and methods in SYSTEMS ANALYSIS.

Simulation

A technique by which a SYSTEM can be analyzed before actual implementation.

Software

The organized INFORMATION used to make HARDWARE functional.

Subsystem

A small SYSTEM placed in a galaxy of larger systems.

Supersystem

A large SYSTEM composed of many smaller systems, i.e., systems, SUBSYSTEMS, and subsubsystems.

Suprasystem

A large SYSTEM composed of many smaller systems, i.e., SUPERSYSTEMS, systems, and SUBSYSTEMS.

System

A multiplicity of parts, elements, or components that interact with one another and work together for some common purpose.

Systems analysis

A process for examining a SYSTEM.

General Bibliography

ALIOTO, ROBERT F., and JUNGHERR, J. A. *Operational PPBS for Education*. New York: Harper & Row, 1971. The concepts and materials presented in this text were written for persons considering or directly involved in the implementation of a PPB system. The book is well developed and easy to read. It includes a large number of illustrations in the appendix, which are appropriate for showing the kinds of forms that can be used in the installation processes. Basic to the text are answers to the "How to do" question.

BANGHART, FRANK W., and TRULL, ALBERT, JR. *Educational Planning*. New York: Macmillan, 1973. This book is divided into seven sections, each of which discusses a component of a seven-element planning model. Each section of the book presents system techniques for implementing the element of the model being described. The book should be of great interest to educational planners and students of educational planning.

CLELAND, DAVID I., and KING, WILLIAM R. *Management: A Systems Approach*. New York: McGraw-Hill, 1972. The objective of applying systems concepts to the management process is the central focus of this book. It is devoid of institutional framework, dealing with management in the generic sense. There are many excellent illustrations and examples. Topics covered include models, the systems approach, planning, organizing, control, and information systems.

———. *Systems Analysis and Project Management*. New York: McGraw-Hill, 1968. In this book, "modern" ideas of systems analysis and project management are presented in a manner appropriate to education. It should have great appeal to those directly involved in decision-making, and provides a good basis for opening channels of communication between systems analysts and administrators. Specifically, the book outlines basic systems concepts. It contains a discussion of strategies involved in the decision-making process. Finally, the authors discuss the execution of decisions as related to various economic, societal, political, and human aspects of project management.

CURTIS, WILLIAM H. *Educational Resources Management System*. Chicago: Research Corporation of the Association of School Business Officials, 1971. This book represents the report of a three-year study by the Research Corporation of the Association of School Business Officials. The study endeavored "to develop a conceptual design for an integrated system of planning-programming-budgeting and evaluating (PPBES) which is appropriate for local school districts." The book provides an extensive discussion of each of the concepts incorporated in PPBES and discusses staff development, federal-state-local interface, and cost/effectiveness.

FRENCH, WENDELL, and BELL, CECIL H., JR. *Organization Development*. Englewood Cliffs, N.J.: Prentice-Hall, 1973. According to the authors of this book, "it is possible for the people within an organization to manage collaboratively the culture of that organization in such a way that the goals and purposes of the organization are attained at the same time that human values of individuals within the organization are furthered" (p. xiii).

The book is easily read, and especially worthwhile for those engaged in systems management. Relevant systems concepts are integrated into the text.

HANDY, H. W., and HUSSAIN, K. M. *Network Analysis for Educational Management*. Englewood Cliffs, N.J.: Prentice-Hall, 1969. The central focus of this book is the development of networks appropriate for CPM and PERT. The authors also provide a number of applications of CPM and PERT to such activities as curriculum development, planning school facilities, campaigns for bond issues, school maintenance, and research.

HUSSAIN, KHATEEB, M. *Development of Information Systems for Education*. Englewood Cliffs, N.J.: Prentice-Hall, 1973. This book is primarily for education administrators. The author endeavors to present the logic and methodology of information-system development. Excellent, easy-to-understand discussions of the analytical techniques and basic concepts necessary in the development of an information system are provided. The book could be appropriately labeled "A Text for the User of MIS."

LEVIN, RICHARD I., and KIRKPATRICK, C. A. *Quantitative Approaches to Management*. New York: McGraw-Hill, 1971. This book is a good basic reference for quantitative methods. A minimum mathematical background of high school algebra is sufficient for the reader. The authors do not find it necessary to use the symbolic notation of advanced texts, nor do they offer rigorous mathematical proofs. Many examples are worked in the book.

Break-even analysis, probability, matrix algebra, linear programming, and queuing are covered in sufficient detail for the beginning users of such techniques.

MCGRATH, J. H. *Planning Systems for School Executives*. Scranton: Intext Educational Publishers, 1972. This book focuses upon "a means for systematic examination of the interrelationships and interactions in the multivariate organizations of public education." Although written for the school executive, it is appropriate for all persons involved in education. Specifically, the book gives attention to: (1) components, subsystems, and relevant variables of school administration; (2) systematic planning and decision-making; (3) awareness of potential of systems design for improving education; and (4) analysis techniques for educational programs.

MODER, JOSEPH J., and PHILIPS, CECIL R. *Project Management with CPM And PERT*. New York: Van Nostrand Reinhold, 1970. This book is divided into two parts, the first of which is introductory in nature and sufficient for most educational users. The material in the second part requires some background in statistics and linear programming. Chapters 1–6 and 11 are all that are necessary for the beginning CPM-PERT user. The nomenclature is consistent with industrial and military standards, and the illustrations are of good quality.

RADCLIFF, BYRON M.; KAWAL, DONALD E.; and STEPHENSON, RALPH J. *Critical Path Method*. Chicago: Cahners Publishing Co., 1967. Although this book was written for persons in the construction industry, it provides a thorough discussion of CPM and is written in language that can be easily understood by educators. The book considers all aspects of CPM as well as the use of the computer in CPM, the role of management, and techniques of project monitoring.

SABINE, CRETA D., ed. *Accountability: Systems Planning in Education*. Homewood, Ill.: ETC Publications, 1973. This book describes ways in which management systems can be implemented in a multiplicity of educational settings. It is regarded as a good "how-to-do" text.

SANDERS, DONALD H. *Computers in Business*. New York: McGraw-Hill, 1972. The objectives of this text are to "(1) provide a general orientation to the stored program computer, what it is, what it can and cannot do, and how it operates —and (2) provide an insight into the broad impact that computers have had, are having, and may be expected to have on managers and on the environment in which managers work" (p. xvi).

The historical development of computers, how they are useful to man, and how they operate are described. No high level of mathematics is used. The application of computers in the areas of planning and decision-making, organizing, staffing, controlling, and economic considerations is covered in detail.

VAN DUSSELDORP, RALPH A., et al. *Educational Decision-making through Operations Research*. Boston: Allyn & Bacon, 1971. This book focuses on the methodology and tools of systems analysis and operations research. Step-by-step introductions and applications are shown. Topics covered include input-output analysis, linear programming, queuing theory, PERT/CPM, and PPBS. The minimum mathematical skill required is simple algebra.